THE WHALE

AND THE

SUPERCOMPUTER

CHARLES WOHLFORTH

THE WHALE

AND THE

SUPERCOMPUTER

■

On the Northern Front of

Climate Change

NORTH POINT PRESS

A division of Farrar, Straus and Giroux ■ New York

North Point Press
A division of Farrar, Straus and Giroux
19 Union Square West, New York 10003

Distributed in Canada by Douglas & McIntyre Ltd.
Printed in the United States of America
Published in 2004 by North Point Press
First paperback edition, 2005

The Library of Congress has cataloged the hardcover edition as follows:
Wohlforth, Charles P.
The whale and the supercomputer : on the northern front of climate change / Charles Wohlforth.— 1st ed.
p. cm.
Includes index.
ISBN 0-86547-659-4 (hardcover : alk. paper)
1. Iñupiat—Fishing. 2. Iñupiat—Social conditions. 3. Whaling—Arctic regions. 4. Indigenous peoples—Ecology—Arctic regions. 5. Human beings—Effect of climate on—Arctic regions. 6. Climatology—Arctic regions—Mathematical models. 7. Climatic changes—Arctic regions—Research. 8. Global temperature changes—Arctic regions—Research. 9. Sea ice—Arctic regions—Research. 10. Albedo. 11. Arctic regions—Environmental conditions. I. Title.

E99.E7W777 2004
305.897'12—dc22

2003019448

Paperback ISBN-13: 978-0-86547-714-8
Paperback ISBN-10: 0-86547-714-0

Designed by Abby Kagan

www.fsgbooks.com

1 3 5 7 9 10 8 6 4 2

To my parents, Eric and Caroline Wohlforth:

along with everything else,

they gave me Alaska.

Contents

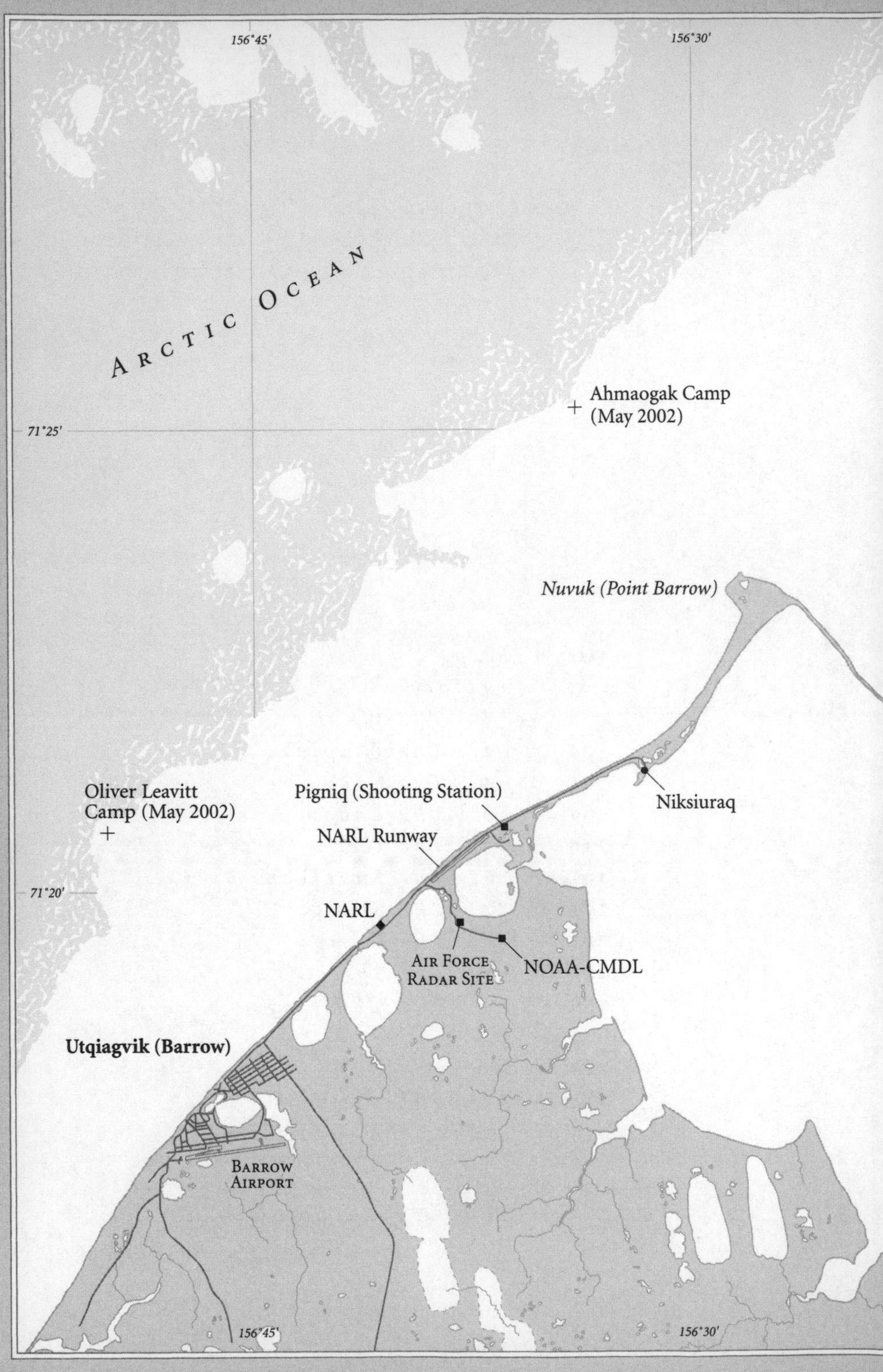
156°45′
156°30′
Arctic Ocean
Ahmaogak Camp
(May 2002)
71°25′
Nuvuk (Point Barrow)
Oliver Leavitt
Camp (May 2002)
Pigniq (Shooting Station)
Niksiuraq
NARL Runway
71°20′
NARL
Air Force
Radar Site
NOAA-CMDL
Utqiagvik (Barrow)
Barrow
Airport
156°45′
156°30′

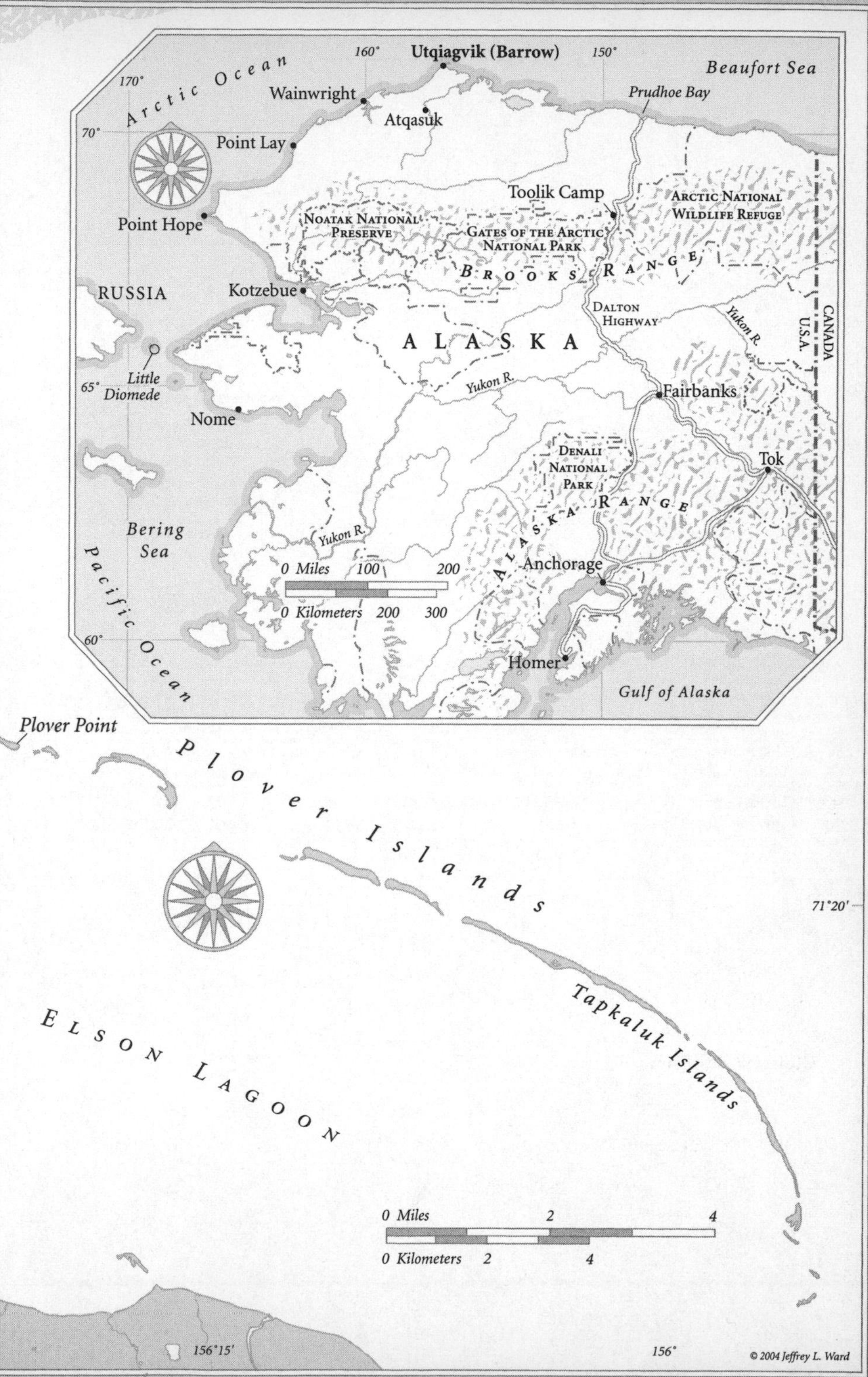

Utqiagvik (Barrow)
160°
150°
170°
70°
65°
60°
Arctic Ocean
Beaufort Sea
Prudhoe Bay
Wainwright
Atqasuk
Point Lay
Point Hope
Toolik Camp
Arctic National Wildlife Refuge
Noatak National Preserve
Gates of the Arctic National Park
Brooks Range
RUSSIA
Kotzebue
Dalton Highway
Yukon R.
CANADA
U.S.A.
ALASKA
Little Diomede
Nome
Fairbanks
Denali National Park
Tok
Alaska Range
Bering Sea
Pacific Ocean
0 Miles 100 200
0 Kilometers 200 300
Anchorage
Homer
Gulf of Alaska
Plover Point
Plover Islands
71°20′
Tapkaluk Islands
Elson Lagoon
0 Miles 2 4
0 Kilometers 2 4
156°15′
156°
© 2004 Jeffrey L. Ward

Preface

I LOVE WINTER. It's when I fly through the birch forest like a hawk. If the snow is good at Anchorage's Kincaid Park, the cross-country ski trails swoop among old trees and over steep, round hills, unwrapping silent white glades and black thickets etched with hoarfrost in quick, smoothly evolving succession. The air feels cool on my perspiring face and steam rises from my chest. Topping a tall hill, I can see gray-blue ice gliding swiftly to sea in the currents of Cook Inlet, the bluffs and low forest beyond, and, on the horizon, sharp-carved mountains, glowing yellow in the low-angle sunshine. Then I push off and hear the wind rushing past my ears as I crouch on the fast downhill. This is what I think of when I'm trapped in muddy traffic in April or when I'm stuck at my computer watching the rain pour down in September. Winter—freedom, purity, grace—the season when the world turns solid, clean, and sharp.

But some recent winters were stillborn in this part of Alaska. Fall came late. At Halloween, when it should be deep snow, we took the children trick-or-treating without coats. The winter's first snowfall was later than ever before, then we had rain and thaw. The ski trails were ruined; running instead, plodding and earthbound, was no substitute. In late winter, normally the best season, the sled dog races were canceled for lack of snow. That almost never happened when I was a child, but now it was happening every couple of years. Some rivers never froze over the winter. Native elders said they had never seen such warm conditions. Everyone talked about it every day, and then everyone stopped. After a while, you couldn't talk about it anymore. Lovers of winter—skiing friends and skaters, snowmachiners, hunters, and dog mushers—all looked stricken and heartsick, and there was nothing left to say.

Science tells us no single winter can be blamed on global climate change. Weather naturally varies from year to year, while climate represents a broad span of time and space beyond our immediate perception. But now science, too, took notice. Average winter temperatures in Interior Alaska had risen 7 degrees F since the 1950s. Annual precipitation increased by 30 percent from 1968 to 1990. Alaska glaciers were shrinking, permanently frozen ground was melting, spring was earlier, and Arctic sea ice was thinner and less extensive than ever before measured. Winter was going to hell.

The Iñupiaq elders of the Arctic noticed first. Sustained for a thousand years by hunting whales from the floating ice, they had developed fine perception of the natural systems around them. Scientists predicted that global climate change would come first and strongest in the Arctic and went there to learn how the sky, ice, snow, water, and tundra interacted to drive changes in the world's environment. Fascinating discoveries accumulated along that path. But the Iñupiat already knew the patterns in the system and how they changed through time, a sense of the whole the wisest researchers recognized and envied. Some sought access to that culture and way of seeing. Others studied how the Iñupiat were adapting to the new world, knowing that the rest of mankind would eventually follow.

The climate here was changing; that was beyond debate. Burning fossil fuels had greatly elevated the carbon dioxide content in the atmosphere. The physics of carbon dioxide trapping the sun's heat on earth, and the rough magnitude of that effect on the planet's heat balance, had been firmly established more than thirty years earlier. We had crime scene, victim, suspect, motive, opportunity, and smoking gun. There was plenty of evidence to convict. We lacked scientific proof to say how much climate change was man-made and how much was natural or to predict exactly what would happen next. The earth is complex; perhaps predicting the future isn't possible. Still, argument raged on over these marginal uncertainties in the face of this enormous, palpable reality.

Let others parry and thrust with the skeptics' abstractions. Here, instead, is climate change in the flesh, the story of individual people at their particular time and place, and what they saw with their eyes and felt in their bones. Here is climate change being lived, the adventure of surviving and thriving as human organisms who must adapt to a new natural world. The Iñupiat have a creation myth about when the earth was upside down; they've been through this before. Christians have their own creation myth; all people have spiritual ideas about land and wilderness. As the world turns upside down again, our species is embarking on an epic physical, moral, and cultural journey. If we're honest, we'll be forced to readjust our fundamental beliefs about how we relate to nature as a species in an ecological niche. The Iñupiat are at the lead, and they seem to be excellent guides.

Over the span of a warm and dreary winter in Anchorage, I learned to enjoy running. When buds formed on the birch trees in time for my father's birthday in April—they used to come nearer my birthday in May—I could only greet them with joy. Day by day, one season at a time, I began to adjust. I was not ready to accept in my heart that the world would always be different, but I was learning to live in the conditions that nature brought to me.

A NOTE ON NAMES

The indigenous people in this book are the Iñupiat (singular or as an adjective, Iñupiaq, which is also the name of their language). They are Eskimos, a name they prefer to Inuit, which is used farther east, in Canada. In Alaska, two major peoples are Eskimo, the Iñupiat of the Arctic Ocean coast and the Yup'ik, to the southwest. In Alaska, the word *Native* is capitalized to mean any indigenous person, whether Eskimo or Indian. The Iñupiat do not use the word *tribe*, but villages are somewhat analogous to tribes, each with its own dialect and customs.

A NOTE ON SPELLING

I apologize to Iñupiaq speakers for transliterating some of the letters in their language that are unfamiliar to most English-speaking readers. We have rendered the *n* with a tail as *ng*. *L* with a dot and *g* with a dot are printed simply as *l* and *g*. In the case of names, I have retained outdated Iñupiaq orthography where that is the preference of those named. No other changes to Iñupiaq spelling were necessary.

THE WHALE

AND THE

SUPERCOMPUTER

CHAPTER ONE

The Whale

THE BRINK OF THE SHOREFAST SEA ICE cut the water like the edge of a swimming pool. A white canvas tent, several snow-machines and big wooden sleds, and a sealskin *umiaq* whale boat waited like poolside furniture on the blue-white surface of the ice. Gentle puffs rippled the open water a foot or two below, except near the edge, where a fragile skin of new ice stilled the surface. Sun in the north reached from the far side of the lead, backlighting the water and picking out the imperfections in this clear, newborn ice with a contrast of yellow-orange and royal blue. This was after midnight on May 6, 2002, three miles off-shore from the NAPA auto parts store in Barrow, Alaska.

A hushed voice urged me on toward the edge.

"Come on, there's a fox. They follow the polar bears."

The fox ran past the camp, beyond the ice edge, danced as it ran, upon that new skin of ice floating on the indigo water. An hour or two earlier there had been no ice there at all and now it looked no thicker than a crust of bread. The fox used tiny, rapid steps. Its feet disappeared in motion. Its back arched high and its tail pulled up tall, as if strings were helping sus-pend it on that insubstantial film of hardened water. Somehow it knew how much weight a brand-new sheen of ice could hold, and knew how to calibrate each step within that limit. The Iñupiaq whalers of Oliver Leavitt Crew watched and muttered with admiration as the fox pranced out of sight. All were experienced hunters, even the young ones, but they were impressed by this skill. This animal knew something valuable, something they would like to know, something that could help them survive.

The five younger members of the crew had been building an ice trail back from the edge. Swinging ice axes and a pick ax, they pounded through ridges as high as a garden shed, pitching broken ice boulders to fill in the dips. The road would be an alternate escape route for the snowmachines and sleds should ice conditions deteriorate, and also a secondary access route to send a young guy back to town for pop and doughnuts if life in camp continued as normal. The young men had been working for twelve hours. Billy Jens Leavitt, the captain's son, was the boss of this job. He was gigantic, tall with huge limbs and feet, swinging a heavy pick ax like a nightstick. His father bragged, by complaining, that Billy Jens tended to throw the harpoon too hard, embedding not only its head but also the shaft in the whale. Ambrose Leavitt and Gilford Mongoyak were Billy Jens's juniors on the crew, but both adults. Ambrose missed his baby; Oliver wouldn't send him home on errands for fear he wouldn't come back. Gilford talked a lot of his one- and two-year-olds. Polite Jens Hopson was a high school kid and Brian Ahkiviana a seventh grader, shy but cheerful, and big for his age. Both had soft young faces but worked like men.

I was older than any of them and they treated me with noticeable respect, and that was a little awkward, since I knew a small fraction of what the youngest of them knew about what we were doing. When I arrived I had to take an ice ax from someone. There were only five axes and six of us. Billy Jens wouldn't tell me whose ice ax to take; he wouldn't tell an older man what to do. They stood around me in a circle as I tried quickly to size up the situation. Then I stepped up and took Billy Jens's big pick ax, thinking that would show I knew he was the boss, and I said, "You look like you could use a rest." In fact, he didn't need a rest, and he liked the heavy pick best. As we started working he took an ax from one of the younger guys, and when I set down the big one for a break he grabbed it but never said a word.

Oliver Leavitt himself sat on a long wooden sled next to his thermos and his VHF marine radio, silently gazing on the water and the ice chunks and bergs drifting by imperceptibly slowly on the calm surface. When I first went out on the sea ice with Iñupiaq hunters I was confused and somewhat bored by long stops when, standing like statues, they stared at the horizon. I secretly thought these guys shouldn't smoke so much if they needed this much rest. One day I learned the purpose of the stillness. I was alone for a while at a whale lookout, pacing for warmth, when a hunter came to my side and took up that gazing posture, as if posing for a romantic painting of a noble Eskimo. Within a minute he pointed out a large polar bear that was approaching about a hundred yards away. To my eye, the bear's appearance was like magic, as if this hunter knew how to summon ghosts from their hiding places. Silent, motionless watching had made the bear visible and prevented us from being potential prey. The whiteness around us, which looked like a vast wreck, a static chaos without scale or reference, in fact was full of information for those who knew how to read it. But first, one must establish a pace slower than the change one wished to observe.

A polar bear swimming past the Oliver Leavitt Camp stopped and paddled in place, raising its long neck far above the water like a periscope to scan the area. On the horizon, across the wide lead of open water, the white tips of jagged pressure ridges showed like the tips of a mountain range on a distant continent.

No longer able to stay awake, I went to join the young guys in the tent. Like all Iñupiaq whalers, the crew used white canvas wall tents, smaller versions of the classic army tent, with sturdy lumber supports and panels of insulated plywood on the floor. A propane burner often brewed a pot of cowboy coffee (gritty coffee made by throwing grounds in with the water), but even when nothing was cooking the flame always stayed lit to keep the tent warm. The plywood grub box contained a bonanza of cookies, candy, and the Eskimos' favorite, frosted doughnuts. Warm meals arrived from home in plastic Igloo coolers: fried chicken, or *aluuttigaaq* (a delicious caribou stir-fry with thick gravy), or a treat of *maktak*, the whale's blubber and skin, raw or pickled. Anything with a lot of fat to keep you warm in cold weather. Next to the grub box were cases of Coke and 7-Up; the ice underneath kept them cool. Socks, gloves, and boot liners hung to dry on the ridgepole of the tent, but everyone had to sleep fully dressed in parkas, snow pants, and Arctic boots. Escaping breaking ice could depend on it; quick escapes happened several times a season. The men, as many as six or eight at a time, slept side by side on piles of blankets and the pelts of caribou and polar bear in an area the size of a king-sized bed. With sleep in short supply, close contact with other unwashed men was no barrier to drifting off.

I woke at 5:30 a.m. to see more polar bears; this time a mother and cub were swimming by, the cub resting on the mother's back. Oliver was still sitting in the same place, looking out in the same direction. The ice continent across the water was closer now: the pressure-ridge mountains were entirely visible. Oliver invited me to sit, drink coffee, and talk. I had been told to keep quiet in whale camp. Crewmen in the whale camp and skin boat, the umiaq, should be quiet and harmonious. Bowhead whales could hear at a great distance and had been seen to divert their paths at a camp noise such as a slamming grub box. In the dark of winter, before the whales arrived, the women who sewed the *ugruk* (bearded seal) cover for the umiaq worked in calm and harmony. When a whaling season went badly, people often said it was because of some conflict going on in town. The Iñupiat dislike conflict. In whale camp, teenagers didn't speak until spoken to. But Oliver said, "Hell, you're not a kid."

Oliver was a big, round man who used his face to tell you where you stood: he could switch quickly from a blank, inscrutable face, to an aggressive "just try me" face, to a knowing smile suggesting you could see half his cards but probably not the best ones. When he was a boy he shot ducks for elders who could no longer hunt for themselves. This skill gave him a small role in an event that helped start the militant phase of the movement for Alaska Native land claims. In May 1961, shortly after Alaska became a state, a game warden arrested a Barrow subsistence hunter for killing a duck out of season. A law made up far away, for reasons irrelevant to feeding Iñupiat families, closed duck hunting from March 10 to September 1, virtually the entire period migratory birds spent in the Arctic. Barrow villagers protested by holding a "duck-in"; they presented themselves to the game warden for arrest, each with a dead duck in hand, almost 150 men, women, and children in all. Oliver's crew provided most of the ducks, passing out about 150 of the 300 they had recently shot, so more people could turn themselves in (some took two, one for the arrest and one for dinner).

Oliver first went to whale camp when he was in fifth grade. His father, who did odd jobs and unloaded freight, didn't have the wealth to outfit a whaling crew, so Oliver went with his uncles and learned from them in the traditional way, by watching, then doing, and receiving sharp correction for errors. One of his uncles would hit crewmen with a paddle; another was kindly—dry Iñupiaq humor can be more corrective than violence.

Starting in eighth grade Oliver went away to a boarding school for Alaska Natives, graduated in 1963, received some vocational training, and lived in New York, Los Angeles, and the San Francisco Bay Area. He served in Vietnam during the Tet Offensive, then returned to Barrow in 1970. An oil company had made a huge find on the North Slope and Alaska Native claims were nearing approval in Congress. The Iñupiat would soon be rich and they needed the help of young men like Oliver who had seen the world.

Sitting on the sled, Oliver was looking for whales and gauging the ice. In traditional spring whaling, the umiaq perches on the ice edge ready for launch. If a whale surfaces nearby, the crew launches as quickly and quietly as possible and paddles to the whale or to a spot where the captain expects the whale to resurface. For the harpooner to hit the whale's vulnerable spot, just behind the skull, with a harpoon made from a long pole of heavy lumber, the captain has to maneuver the boat right onto the whale's back or within touching distance alongside. The whale can move much faster than the boat, so most of whaling is waiting quietly for a whale to come close enough to launch. In camp, no bright colors are allowed that might catch a whale's eye and crews avoid unnecessary noise and movement. Hunters wear white pullover parkas lined with caribou hide for camouflage on the white snow. That morning we saw only one whale, a far-off black back rolling across the surface, and heard another, a roaring blowhole exhalation from somewhere we could not see, hidden by the ice. Normally at this time of year, a crew would be seeing whales every few minutes. Crews farther down the lead were paddling in search of one, thinking the migration might be passing by on the other side of the big ice across the lead.

The ice was bad that year. It had been bad for a decade and seemed to be growing steadily worse.

The shore ice should form in the fall as bergs left over from the previous year float near the beach and are sewn together by new ice that freezes in the cooling temperatures. These big bergs are chunks of the previous year's ice pack that never melted over the summer. They usually form out of old pressure ridges, mountains of ice built by the collisions of huge ice sheets, becoming freshwater ice as warm spring temperatures drain pockets of brine trapped inside. The surface becomes rounded and smooth and the ice becomes dense, hard, and brittle. The Iñupiat call it *pigaluyak*, or glacier ice. Under the surface whiteness it glows iridescent blue, like a glacier. Iñupiaq travelers use the fresh water for making tea far from home. Whalers seek out multiyear ice; it provides a strong platform for pulling up whales and it anchors the shorefast ice in place with its great mass.

In the winter of 2001–02, however, as for several years prior, little multiyear ice had appeared at Barrow. The shore ice didn't form as solidly as it should, and it lacked the big, solid anchors that multiyear ice, or even new ice with large pressure ridges, would have provided. And on March 18, something strange and unsettling had happened. The ice went out, leaving open water right up to the beach in front of Oliver Leavitt's house. A distant storm had created a tidal surge near Barrow that lifted and cracked the ice pack; a current had pulled it away. (The Arctic Ocean has virtually no lunar tides, but atmospheric pressure gradients cause rising and falling water levels, which the Iñupiat call tides). The ice should have been strong enough to withstand that. No one could ever remember the ice going out that early. Normally, it goes out in July. A dozen seal hunters floated out to sea on the ice. The North Slope Borough's Search and Rescue helicopters went to find them and bring them home. Some didn't know they were drifting away into the Arctic Ocean until the helicopter showed up for the rescue. You can't tell you're moving when your whole world starts to drift away.

Later, ice returned and refroze to the shore, but it wasn't sturdy ice and it still lacked good anchors. As whaling season began, a strong west wind pushed the ice against the shore for several days, then a strong east wind tested it and cleared away some of the junk ice. Oliver's theory now was that these events had cemented the ice adequately for safe whaling. He had chosen a flat area of ice with a color and height above the water that told him it was strong enough to pull up a whale. But every so often he sent someone to look at the watery crack that was a little behind us or to check the dark ice—weak, brand-new ice—that lay a mile or two back, between us and dry land.

Another threat occupied his mind even more that morning: the big mass of ice we could see across the lead, which was moving very slowly toward the southwest but also seemed to be getting closer at an imperceptible pace. Oliver said, "That's the dangerous ice. If people start noticing it's coming in, we'll be out of here in five minutes flat."

The momentum behind an ice floe, even if it is moving only slowly, is stupendous; when it hits the unmoving shore ice, the collision is like an immense, mountain-building earthquake, a terrifying event called an *ivu.* Oliver was young at the time of the big ivu in 1957, but he remembered how the ice went crazy, with big multiyear floes standing up on end and shattering far from the edge, forcing the crews to scramble for their lives over miles of cracking, piling ice, leaving camps, boats, and dog teams behind—their entire means of supporting their families. Now he told the story of an ivu in the 1970s that came while the village was butchering a whale caught by his uncle, Jonathan Aiken. Men raced away on snowmachines, dragging off huge strips of maktak from the whale's side even as the ice devoured the whale's body. He told stories like this frequently in camp while the younger members of the crew listened. The ivu stories were the scariest.

In 1978, an elder, Vincent Nageak, told gathered villagers the story of an ivu in which boats and dogs were lost, as was one man, Aanga, who was caught by a moving chunk of ice:

> Right after it had bit Aanga in its grip, they tried to hurriedly . . . remove him from there all right, when [the ice] stopped for a little while, but he told them, "I don't think you can take me off from here with those little penknives, do you?" And here he was with his pipe in his mouth, they say. "I don't think you can be able to take me off with those little penknives, do you?" And so immediately after he had finished saying that, all of a sudden, without warning, it began again, and so Aanga [was taken] down under. Holding his pipe in his mouth, it is said, after [the ice] had bit him in its grip, when we was about to go out of sight, he just smiled at those [who were] there.

Iñupiaq chatter on the marine VHF radio by Oliver's side began to flow with comments from nervous captains up and down the lead. They saw the big pressure ridges across the open water growing noticeably closer.

The radio box was a sturdy plywood case with a car battery and a tall boat antenna. Each camp could hear its "base," usually the captain's home, and other camps spread out along twenty miles of the lead. Channel 72 was for whaling and channel 68 for routine in-town communication. VHF sets seemed always close at hand near kitchen tables and under the dash of pickup trucks. In the morning people said "Good morning, good morning" to announce they were on the air, and the NAPA auto parts store—which carried harpoon parts and other whaling supplies—let everyone know when they were open for business. In the evening, each person said "Good night" when turning off the radio, and the children and grandchildren of whalers said good night to their fathers and grandfathers out in camp—sweet broadcasts of kisses and love names that the whole town could hear. One evening I heard tough old Oliver Leavitt on the VHF trading silly I-love-yous with his granddaughters, Ashley, age seven, and Appa, four. During times of peril, the VHF allowed whalers to act almost as one, sharing observations about ice and water movement and dynamics from many perspectives.

The whalers handled these technical conversations in Iñupiaq, even though many younger people are not fluent. Some handy words don't exist in English, such as *mauragaq*, to cross open water by jumping from one piece of ice to the next; or *tuagilaaq*, to kill a whale with a single blow to the sensitive spot behind the skull; or *uit*—literally "to open one's eyes"—a term used to indicate the breaking away of pack ice from shore ice to form an open lead of water. But one cannot attain the full benefit of Iñupiaq by simply incorporating individual words into English as technical jargon. The very structure of Iñupiaq helps deal with situations in a unique environment. Speakers can convey information quickly in a moving landscape without landmarks or any visible distinction between ocean and shore. In the absence of physical reference points, the speaker can position objects and events using movement, the relative locations of speaker and listener, and the directional orientation of the ocean and rivers. For example, *pigña* indicates that the thing you are talking about is above, has a length less than three times its width, is visible and stationary, and stands at equal distance between speaker and listener. *Pagña* contains all the same information, except that the subject's length is more than three times its width. English has a few such words, such as *hither* and *yonder*, but they are largely obsolete and not nearly as useful. Iñupiaq endings also aid coordination by allowing speakers to pass on oral information without losing nuances about the quality of the knowledge and how it was obtained. They cover a gradient roughly ranging from "I saw it myself and it is certain" to "Someone saw it and it might be true." (Contrary to popular belief, however, the Eskimos do not have a hundred words for snow.)

From Oliver's vantage he could see that the apparent nearing of the ice across the lead was only the passage of a point; part of the ice island drifting parallel to our ice protruded in our direction, creating the illusion that the entire floe was moving quickly toward us. Oliver uttered a few words of Iñupiaq on the radio and the discussion stopped. "You got to talk to them quick before they scare themselves," he said. Each captain's experience and expertise were well known, another factor in how whalers evaluated conditions and safety. Oliver Leavitt's name carried unquestioned authority.

When I first saw Oliver Leavitt Crew at work they were building the boat that now stood at the edge of the ice. The Iñupiat Heritage Center, a well-equipped cultural center and living museum in Barrow, had a large workshop called the Traditional Room, where whalers, artists, and others involved in cultural activities came to build things. Oliver's boat was on one side while Julius and Delbert Rexford's *Atqaan* crew repaired and mounted a new skin on their umiaq in the other. Oliver was known for his boat-building skills. He worked with a few crew members his own age, men like Hubert Hopson, strong and skilled but past their prime, giving them instructions as equals. The next generation in the room was represented by the Rexfords and their senior crewmen: they were around my age, nearing forty, with plenty of responsibilities of their own—Delbert was a former borough assembly member. But they asked for Oliver's opinions, and when Oliver saw Julius doing something he didn't like—attaching a piece to one of his own sleds—Oliver gruffly told him so and Julius quickly changed it to the way Oliver recommended. Next in seniority were Oliver's younger crewmen, men in their twenties, including Billy Jens, Gilford, and Ambrose. They did skilled work, but under direct supervision. Oliver taught them and they listened carefully. At the bottom rung, teenage boys stood around the edges of the room waiting to be told what to do and holding their tongues. They wore their snow pants and warm boots indoors, aware that they could be sent to the store or the wood pile at any moment and would have to jump quickly.

I once asked a high school class in Barrow why teenagers were so respectful around their whaling captains. They politely made it clear this was a foolish question: no one in his right mind would risk his place on a whaling crew, a position of high status for even the lowest member. They would as soon risk a chance for a basketball scholarship. (Barrow's other mania is basketball. Besides closely following the teams of the NBA and

NCAA, young fans knew the name of the first Iñupiat to dunk the ball, a player for the Barrow High School Whalers.) But fear of being excluded from whaling was only a part of it. Young people in Barrow generally were more respectful than they were in Anchorage. Respect is fundamental in Iñupiaq culture; I think they learned it by seeing it practiced.

As the day wore on in camp, the young men did chores, checked the ice, and built a blind of ice blocks. Jens tried to complete an American history term paper that was due in a couple of days. Oliver informed him about President Warren G. Harding and the National Petroleum Reserve at Teapot Dome, Wyoming. He then told the history of the National Petroleum Reserve–Alaska, just south of us, where oil exploration brought modernity to Barrow fifty years ago and where an oil strike and the new money it would bring were still hoped for. Oliver claimed he wasn't as articulate in English as in Iñupiaq, but he knew how to tell a good story: slowly, with sharp, percussive words, well-chosen profanity, a clear point at the end, and often a punch line.

Many of the stories were about whaling screwups: mishandled weapons, misread whale behavior, missed signs of danger on the ice. Some stories taught the history of grappling with the white world. Oliver had presided at a meeting in the early 1970s when the oil companies sent a chartered 737 full of lawyers to Barrow to say they didn't recognize the new local government and wouldn't pay property taxes on their oil discoveries at Prudhoe Bay. Oliver and his colleagues got the upper hand through sheer terror, playing the part of the exotic primitives. They carried lawyers in business suits to the meeting on long snowmachine rides in bitter cold March weather (really, it was a short walk). Oliver used a billy club for a gavel and threatened to mace the audience if they didn't maintain order, offering a safety briefing to go along with the threat: "The trick is don't panic. You will be taken to the hospital. The doctors will wash your eyes out. There's no running water here. Just don't rub your eyes."

As afternoon progressed, the sun was bright and unseasonably warm. The ice reflected brilliantly while the deep, dark water swallowed light. The details of the pressure ridge mountains across the lead were clearly visible. The radio grew lively again. Oliver stood and watched the ice across the water intently. Everyone else stood, too, waiting for what he would say. Then, calmly, "We better start packing up."

The younger men began by emptying the tent. Oliver and Hubert worked on disabling the weapons and putting away the radio. Now you could see the ice moving through the water directly toward us. The work already was going fast—everyone knew his job without a word—but Oliver said, "Better hurry up, Billy." When speaking to the younger part of the crew, he addressed only Billy Jens, like an officer giving orders to a sergeant. Billy Jens grunted a few orders to the others. I tried to help, grabbing plywood floor panels and dropping them on a sled. Again, "Better hurry up, Billy." Oliver's voice had an increasingly urgent edge to it. Things not fitting in right, the boys started throwing stuff on the sleds haphazardly. Less than ten minutes had passed, but the lead was quite narrow now, just a few hundred feet. "Better hurry up, Billy," the tone each time a little higher.

Billy Jens had too much to do; I tried to tie down one of the sled loads for him, using half-hitch knots for speed rather than the quick-release knots the Eskimos prefer. The load came loose—I had tied the lashing line to the wrong rope. Billy Jens came around to retie it, but my tight half hitches were tough to untie. The boat was ready to go on the sled. The ice was a hundred feet away and closing fast. Oliver said, "Billy, pick up the boat." Billy was still trying to fix my mistake, without saying a word. I tried to help with the boat, but I didn't know where to grab it. "Better hurry up, Billy." Billy Jens grabbed the gunwale, we heaved the boat onto the sled and started strapping it down. I could see the crystal-thin rim of ice where the fox had run the night before and, with the lead of water almost closed, I could see the same rim on the other side approaching. We needed to escape before a possible ivu could break our ice free from the shore. "Better hurry up, Billy." Miscellaneous gear was thrown in the boat. Oliver told me to grab the back of a sled, where I stood, holding the handles. The snowmachines moved into place to hitch up. Only minutes had passed.

As I was jerked into motion behind Hubert's machine, I could see the collision begin. The glassy film of new ice from each side made contact and the delicate tracery that had supported the fox shattered and disappeared into the ocean.

We bounced wildly down the ice road we had built the night before, the boats pitching up to crazy angles on their sleds before they topped the ridges and raced down behind the snowmachines. The trail that had seemed so smooth and straight when we were chopping it now swung me wide with each turn on the back of a heavy sled of gear, crashing into ice blocks and ridges while I held on tight, bracing for each jolt. Hubert turned his head and yelled, "Steer!"

Steer? How?

We stopped on a big flat pan of ice near town. Crew after crew filtered in from the trails and joined us, until rows of sleds and boats stood side by side as if in a big parking lot. It was sunny and warm and a good time for friends to meet—teens with teens, captains with captains—and to talk of guns, snow machines, and ice conditions. No fear, no sense of relief. These days, with the bad ice and warm weather, an escape like this was routine.

John Craighead George, known as Craig, arrived in Barrow in 1977, having taken a job caring for animals at the Naval Arctic Research Laboratory, America's primary research station in the Arctic, universally known as NARL. He had escaped to Alaska from Moose, Wyoming, at the base of the Tetons, a refugee from an illustrious family. His uncles, John and Frank Craighead, were world-famous experts on the grizzly bears of Yellowstone National Park. His mother, Jean Craighead George, was the author of the classic children's books *My Side of the Mountain* and *Julie of the Wolves*. Craig himself was a scruffy mountain climber, a member of a Jackson Hole fraternity dedicated to redundant conquests of black Grand Teton and its pointy neighboring mountains. His grades were lackluster, his scientific experience thin, and his family's success daunting. "That's why I ran," he said. "It's kind of a heavy dose."

A first flight into Barrow felt like flying to another planet. You took off in a jet from Fairbanks, a town of 80,000 in the middle of Alaska, without having left the usual interchangeable mall of gates, jetways, and airplane cabins. On a clear day you could watch untracked wilderness pass below you for a solid hour and a half: the great river valleys of the Interior, the insane multiplicity of mountain peaks in the Brooks Range, and then an area that resembled infinity—the flat green tundra of the North Slope with its randomly wandering rivers and lakes like water droplets on a windshield. On the plane's descent nothing man-made appeared outside the window—no roads or buildings, just more oblong lakes and strands of swampy tundra—and it looked like you were going to land in one of them. The town showed up just before touchdown. From the air it looked like a load of cardboard boxes that fell off a truck in a place identical to every other place as far as the horizon.

On a map, Barrow's location appeared quite logical, at the extreme northern tip of Alaska, on a shoreline that showed up, like any other, as a sharp boundary between the manageable terrestrial kingdom of dry land and the invisible wilderness of the sea. But here that line was a misleading metaphor. A smudgy watercolor gradation would be a more accurate one. Land and ocean were both made of water. During ten weeks of summer, the land, flat and treeless, was only a slightly warmer shade of green than the ocean. A thin layer of tundra lay atop permanently frozen ground called permafrost. Surface water sloshed around on that ice. Innumerable lakes, all lined up southeast to northwest, perpendicular to the prevailing wind, shredded the very idea of land into temporary lacy tendrils that barely stretched around reflections of the sky. The line between this flat, wet ground and the flat sea seemed entirely arbitrary. For nine months of the year the sea and land merged in whiteness. In winter, the visual difference between land and shorefast ice was merely one of texture. The shapes on shore were soft, those on the sea angular. At times, an ivu smudged the distinction further, pushing over the shore, digging and overtopping the land, and crushing anything in its path.

The town itself was strange to a new visitor, too. You left the aviation cocoon abruptly, disembarking not into a terminal but onto cold tarmac. When Craig arrived, the airport terminal consisted of a twenty-foot-square building with blankets hung in the corners to hide the toilets, which were buckets. Even in 2002, Alaska bush communities had no business district, no shops, and few signs (everyone already knew where everything was). Other than NAPA and the building supply store, everything came from Stuaqpaq ("big store," in English), where you could buy groceries, clothing, ivory carvings, all-terrain vehicles, deli sandwiches, books, music, and so on. The houses, arrayed along wide gravel roads, were mostly plywood cubes sitting on pilings—the warmth inside isolated from the permafrost to prevent the ground from melting and turning to mush. Meat or fowl would hang to dry outside, snowmachine parts and animal bones lay here and there, and dogs orbited on tethers. To visit, you had to enter, unbidden, the first room of the house, the *quanitchaq*, or Arctic entry, a sort of airlock to keep the cold weather at bay. For anyone to hear you inside, you had to knock on the inner door, inside the quanitchaq. But in most Eskimo homes, the quanitchaq was piled high with boots, coats, toys, tools, snowmachine oil, garbage, frozen meat, skins, and other miscellany—it seemed like a personal space—and shutting the outer door before knocking on the inner door plunged you into total darkness. Even the addresses were indecipherable to an outsider because, instead of using street names, each building was designated solely by a unique number. To find your way around, you simply had to know the numbers: when a person said "4337," for example, that meant the mayor's house.

I was always treated with politeness and respect in Barrow—although the form politeness sometimes took in Alaska Native culture could come across to whites as sullenness—but for a long time many people couldn't remember who I was from one meeting to the next, either by name or by face. Craig joked to me, "You've got to do your apprenticeship. You've got to be here five years before they'll recognize you, and then another five years before they'll talk to you." My theory was that Native people learned unconsciously to ignore a new white face until the newcomer behaved like a member of the community. From the perspective of traditional Iñupiaq norms of behavior most whites were rude: they talked too fast and didn't give others a chance to say anything, they stared, they spoke too directly, contradicted others, and didn't listen for meaningful nuances, they couldn't sit still, and they didn't reciprocate the gifts of knowledge and hospitality they received. Iñupiaq people spoke slowly, used stories to make points, and always avoided conflict; an elder once expressed a strong disagreement to me by saying, "Different people see things different ways." Besides, outsiders cycled through too fast to remember them all, and local people often got burned by their brief, voracious presence: burned by unethical journalists, researchers, and contractors, and burned by people who came to work just for the high wages or gave up on the place after a year.

And from an outsider's perspective, who could stand to live long in the Arctic on the outside of that chilly shell? When Craig saw how flat the place was and how different his life would be, he freaked out and decided to leave.

His boss persuaded him to give it another chance. When Craig arrived, the foundation of Iñupiaq culture was in peril. Iñupiaq whaling appeared to be in its death throes. In 1977, Eskimo whalers struck 111 bowhead whales out of a population estimated by government scientists to number only 1,300. The International Whaling Commission, a treaty organization dedicated to saving the world's whales from extinction, ordered the hunt stopped. The word from the IWC arrived on the North Slope as a fait accompli. The Iñupiat mourned. Whale accounted for a large percentage of the protein they consumed, but the death was a cultural death. And it seemed unnecessary. The whalers and elders knew there were more whales than the scientists had counted.

The bowhead, also known as a right whale, is a large baleen whale. It filters shrimp and other small creatures through fibers on rows of hundreds of flexible, bony black strips called baleen, which range in length from a few feet to more than fourteen feet. Before the advent of plastic, baleen was a common household material used in anything that needed to be strong and flexible, such as umbrellas and women's corsets. Nineteenth-century commercial whalers depleted the Atlantic right whale to the point of extinction, where it still hovered a century later. The right whales of the Pacific were distinct, with slight physiological differences. They were called bowhead for the hard, bony skull that allowed them to surface through thin or broken ice. Commercial whalers overhunted the bowhead, but not as severely as they did the Atlantic right whales. Eskimo whaling traditions waned but never died out completely, and, by the time the Iñupiat were told to stop, they believed with confidence that whale numbers were strongly rebounding.

What chance was there that the world would believe Eskimo hunters over learned scientists? A paper delivered to an IWC conference by U.S. government scientists predicted eventual extinction of the bowhead even if no hunting were allowed. In the world of science and public policy, Iñupiat knowledge appeared anecdotal and self-interested. Some environmentalists, including one senior IWC official, charged that the Iñupiat's culture was inauthentic because harpoons were tipped with metal instead of bone and said that the whaling captains were really just trophy hunters. Even some allies privately doubted the Eskimos' claims. But, thanks to Prudhoe Bay crude oil, the Iñupiat had the resources to push their case. They had won the fight to tax the richest oil field in North America by forming a county-level local government covering the entire region. The North Slope Borough helped fund creation of the Alaska Eskimo Whaling Commission, which successfully lobbied with the international commission for an annual quota of twelve whales—not enough to feed the nine Iñupiaq whaling villages, but something to keep the tradition alive. Meanwhile, the new borough's Department of Wildlife Management (known around town simply as "Wildlife") hired scientists to develop knowledge in the western style that could support the subsistence way of life.

Tom Albert, a research veterinarian from Pennsylvania, became the senior scientist of Wildlife (established in the facilities of NARL, which was closing) and hired Craig to manage the new whale count program. Harry Brower, Sr., became Tom Albert's teacher. Harry was one of Barrow's most respected elders, a successful whaler and a former science support worker at NARL, and he knew how to communicate his knowledge to Tom and the other scientists. Other whalers trusted Tom and his team because Harry did. Harry died in 1992 and Tom retired back to Pennsylvania in 2001, but their relationship lives on as a legend in Barrow, memorialized by a museum exhibit in the Iñupiat Heritage Center.

Harry and other whalers explained what the scientists were missing and why. The whale counts used the same simple technique used to count whales anywhere: pick a spot along the migration route, count the whales you see, then make a statistical adjustment for the whales you think you missed. The Iñupiat believed this methodology missed whales two ways: the migration passed on a much wider band than the scientists assumed, and the whales often swam under the ice, where they could not be observed. Eskimos could hear them day and night spouting through holes in the ice.

Craig's team would have to invent a way of counting unseen whales and do it out on the dangerous ice. In Barrow's long scientific history, NARL researchers and later those at Wildlife had come to believe that, as Craig said, going out on the ice without a Native guide was "sheer death." Scientists always had appreciated the Iñupiat's superior practical understanding of sea ice, although they often discounted the fundamental knowledge that supported that practical understanding—knowledge of ice strength and dynamics, weather and currents. Craig had all that to learn. He was tall, muscular, an experienced outdoorsman—his climbing experience had gotten him the job—but his outdoor skills came from recreation, not work. Mountain climbing, however adventurous, was an unnecessary hardship. Mountaineers valued an accomplishment for its difficulty. The Eskimos' utilitarian way of camping shared nothing with the ideal of contrived challenges.

"My first trip out with Benny Nageak, camping on the Colville River," Craig recalled, "I pull out this dome tent and a little baby MSR stove and some dried food. And Benny's like, 'Wait a minute. Where's the wall tent? Where's the Coleman?' And you know, he basically said, Where are the cookies and the coffee and all the good stuff? And the pop? And he said, 'For God's sakes,' he said, 'you don't know how to camp.' And here I had just climbed McKinley and all this kind of stuff, and I went, 'Oh my God, I'm not in Kansas anymore.' There were a lot of things like that. I didn't bring a radio. You know, you never brought an AM radio going up in the backcountry in Yellowstone. That would be sacrilege. He was like, 'Where's the radio? The VHF and the AM radio? We can't listen to KBRW and the birthday program. What the hell is this?' "

KBRW was Barrow's radio station and a key node in the communities' neural network of communications. Out on the ice, upriver in fish camp, or in the pickup truck around town, KBRW was often playing, relaying *All Things Considered*, National Public Radio interviews with authors and classical musicians, frequent weather forecasts, the evening show when elders told stories in Iñupiaq, the morning show with interviews of visiting scientists and politicians, and, perhaps most popular of all, the birthday show. Each evening around dinnertime listeners called in to wish a happy birthday or happy anniversary to whomever they wished, short messages following a relaxed kind of ritual form, flowing gracefully one right after the other like a psalm of goodwill. Safety came from being part of the community. No radio meant you were alone.

Craig and his colleagues set themselves to learning Iñupiaq ways of evaluating the ice and keeping safe while they developed technology to substantiate Iñupiaq perceptions of how many whales were underneath the ice. Still, Craig remained skeptical about the whalers' claim that bowhead numbers were good and that the scientific consensus was wrong: "We weren't sitting on a thousand years of traditional knowledge, and we frankly were taught we were scientists and we were doing stuff scientifically, carefully, and the other information was anecdotal." The methods they developed would have to withstand tough peer review from other scientists who felt the same way. The answer would be a number and, underneath that number, more numbers—data points corresponding to whale sightings, processed statistically using methods mathematicians could accept. The only channel for Iñupiat knowledge would be in the help offered to keep the workers alive on the ice and to tell them where to look for whales.

Counting whales you couldn't see wasn't easy. Not until 1984 did the team perfect the gear to do it: arrays of hydrophones that could listen for whale vocalizations and triangulate the animals' positions using the time it took for sound to arrive at widely separated points. Specialists in submarine acoustics, electronics, mathematics, and computers built the equipment and the computer programs to pinpoint individual whales and keep track of them as they slowly swam through the twelve-mile range of the microphones. Craig himself came up with the idea of putting the equipment on enclosed sleds researchers could drag over the whaling trails behind snowmachines. Scientists linked the equipment to synchronize the signals, then matched the sound print to visual sightings made from a series of perches set up on high pressure ridges. It took an army to field the entire array of sensors and eyes, maintain the camp, and guard against polar bears, and to do it around the clock, whenever conditions permitted, for a migration that started in April and ended in June.

The first year with the new equipment was a disaster. The inventions worked—on the headphones it was possible to hear whales passing in stereo—but the ice conditions were bad and the count failed.

"Set up the gear, put up a perch, and slam, the sea ice slams into it, and we'd lose all our gear. All of it," Craig said. "It was like a war. We were using lead acid batteries. Fifty-pound batteries. All our clothes were ruined with battery acid."

Each spring a couple of dozen workers arrived in Barrow for the count, mostly young academics, bunking together in NARL's old Animal Research Facility, known as the ARF. The ARF was low and weather-scarred, like other buildings at NARL, but inside it became the home of two decades of exciting, exotic wildlife research. The walls were papered with homemade cartoons about the count, data sheets with facetious entries alongside graphs of real, hand-plotted data, letters and postcards from former workers, and signs importuning everyone to take off bloody or muddy boots and to unload guns at the door. Whaling season involved days and sometimes weeks of waiting for weather to break. Workers listened to KBRW, swapped paperback books, and talked around the kitchen table. Information flowed through the office, a radio room like those in old war movies.

Bunk beds lined a series of strangely shaped rooms—some of them former animal examination rooms—and the TV and VCR stood in the pantry, with shelves solidly stacked with cans of Spam and Dinty Moore stew. Through a door at the back of the living quarters lay Dr. Frankenstein's lab, where biologists processed and stored their samples—sinks, a Butcher Boy band saw, freezers, a workshop, walrus skulls and whale bones, and shelf after shelf of jars with labels such as "walrus fetus," "bowhead brains," "polar bear head," "whale lice," and, my favorite, "testes, unknown."

Barrow became a center of whale research because only here could scientists get samples of large, freshly killed baleen whales. Wildlife became the gatekeeper for a stream of researchers who wanted a piece of a whale. Working with some of those scientists, Craig made an important discovery. Another scientist had experimented with measuring the aspartic acid (an amino acid) in whales' eyes to find their age but gave up because the ages always came out too high. A test of a newly mature female came up with an age of twenty-three years, which seemed impossible. In 1992, Craig was present at the butchering of a big bowhead when a friend pointed out an ulcer on the skin. He reached his hand into the doughy wound and pulled out a stone spear tip. The next year, a stone spearhead was found in another whale, and there have been several others. The Iñupiat switched to metal harpoons in the 1880s, so these whales had to be more than 110 years old. Using that knowledge, Craig and his colleagues calibrated the aspartic acid aging technique and established that the bowhead's normal life span was around 130 to 150 years, making it the world's longest-living mammal. Results suggested one whale was 211 years old.

The most important project, the count, finally hit its stride in 1985. Using the new equipment and with the help of statistician Judy Zeh, who brought more sophisticated mathematical techniques to accounting for missed whales, the team produced a population estimate of 7,200 whales, almost six times the original midrange estimates. Wildlife also produced evidence showing the dietary needs of the villages. The IWC responded by cautiously increasing the Iñupiat's bowhead hunting quota by a few each year. Counts in subsequent years yielded ever-higher whale numbers. Despite increased Native whaling, the bowhead population was strong and rising. In 2002, Craig missed spring whaling to present findings at the IWC meeting in Japan. The new official number was 10,000. The quota had reached a level the Iñupiat considered adequate: seventy-five strikes, or fifty-one whales landed, whichever comes first, distributed over ten whaling villages with 163 crews.

"The Natives were vindicated," Craig said. "They were right. They were right about all these things."

Like any legend, the story of how the Iñupiat regained whaling became a lesson and a touchstone, both for the Eskimos and for environmental scientists studying the Arctic. From the Native perspective, the whale count amounted to a lot of trouble (and $10 million) to find out something they already knew, and they related the story to demonstrate how Iñupiaq culture could win if its defenders fought hard enough. But from the scientists' perspective, the lesson was about knowledge. Researchers—at least those paying attention to anything beyond their own work—had to accept that there was another valid way of knowing complex facts about the environment. Indeed, for this system of many parts in constant change, the Iñupiat were able to draw broad, useful conclusions in near real time, something hypothesis-based science couldn't come close to doing.

Other controversies that arose made the same point again. The oil industry wanted to operate ships that emitted loud noises to take sea floor seismic readings. The science said the ships would affect whales only up to four miles away. The Iñupiat said they would affect whales much farther than that. Better studies conducted in 1996 to 1998 found whales deflected from their paths twelve miles around the ships, at the very least. In other cases, Eskimos' knowledge improved scientists' surveys of land animals to help set hunting levels. And so on.

No one reverse-engineered Iñupiaq discoveries to find out how the community reached its conclusions, but Craig thought about the question as a scientist respected by both cultures. Iñupiaq skills of observation and communication were the key. Every Native home was an information node. The effect could be disorienting: the VHF competed with KBRW, a pinochle game, a TV, and sometimes a computer, kids ran around, and elder family members sat at the table sipping coffee and telling stories. At the Volunteer Search and Rescue base many of those elements were present, plus a pool game in progress, a hunk of walrus boiling, a big wall map to coordinate rescues, and various different conversations in two languages. Everyone seemed to share information all the time, from around town, out on the tundra or the ice, and around the world. I learned on my first visit to Barrow that if I asked a question someone else might give me the answer the next day—you couldn't say a word you didn't want everyone to hear.

Outdoors, communication among many observers—hunters and fishermen—created a continuous picture of a large area, an area far larger and for a duration far longer than the measurements taken by any scientific team going into the field for a few weeks at a time. The Iñupiat were trained observers, they were covering a whole system for a long time, and they had a way to process the enormous data set they derived into useful information for making decisions. In science, you come closer to truth—known as a high statistical level of confidence—as you add more data points. Craig compared the community to a giant machine gathering and crunching data.

"It's a bit of a black box to me," he said. "There's conversation, conversation, conversation back and forth, and then there's this statement that comes out: 'We know this.' They're taking in massive amounts of data and processing it like a supercomputer."

Arnold Brower, Sr., one of Barrow's most successful whaling captains, now in his eighties, had watched as the Arctic climate changed. "Unusually changed," he said. "And the pattern of animals, as to how they behave, like caribou and the fish, the seasons of spawning and seasons of ice forming on the surface.

"And that blanket of clouds that had been protective of tremendous heat that would be there every day for twenty-four hours. I don't know how it is that it got up there that way. . . . And the temperatures are warmer. Earlier and later, are freezing and getting where, kind of, you can't predict each year to be the same.

"The weather patterns, you can pretty near predict if weather is going to be cold at that time or whether it is going to be warming up. . . . But I think we had a crazy type of a change.

"The current has been kind of unpredictable here, because the current would change and then it would change back, and sometimes it would quiet down and form into like a big pool of water, a lake out there in the ocean. And all of us sat there and without hardly any warning the current would shove out to one side and run for a week and then change over again. But in younger years it used to be two-way, and it would take time, and the wind wouldn't change it at all. . . . So it's not predictable at all what it's going to do next. It's unusual.

"It was something that you could predict to go out there and hunt all day and not think about getting stranded out there, if you use the method of sounding and the wind."

Kenny Toovak, born in 1923, said, "In the springtime, we used to go out camping on the riverbanks and wherever and geese hunt. Sometimes we spend out there a week, ten days, two weeks. And the thaw was kind of gradual. Each time the snow melted in a gradual way. Today, seems like good and hard snow, and then overnight the weather change, the temperature, and the sun gets clear and sunshine, and in one day's time the snow started to melt away. Kind of rapid.

"My parents used to go out camping after the whale season out down the coast, down coast by Will Rogers monument, Ualaqpaa or farther from Ualaqpaa sometimes, we'd spend the summertime out camping in that area or hunting for ugruk, seals, or whatever. You name it. And sometimes our old tents were not that good, that we had for years and years, and sometimes they've always got a hole in part of the tent. But we never get bothered by mosquitoes. Hardly any mosquitoes in that area in the summer months. That means kind of decent temperature. So today, when you go down coast and hit the beach, boy, there are all kinds of flies. You've got to have that insect repellent, boy, all over."

Harry Brower, Jr., the son of Tom Albert's friend, collects traditional knowledge for Wildlife and is a whaling captain. He said, "It's hard to find a place to pull up the whale. If you have this first-year ice, it's not really thick enough to hold the whale, pulling it up out of the water. With that multiyear ice, if you have a large pan identified, you could pull up a whale there. We had an area specifically last spring where one big pan had fastened itself to that first-year ice and it probably was grounded, because of that weight. At the time we had a real high surge of wind and the ocean current or tide went way up and the big ice pan came in, and when the current shifted and the tide went back down I think it got grounded and it stuck there all throughout the late spring, and early summer it finally drifted out. And I think we butchered like seventeen whales on that ice pan. It was the only one heavy enough to pull up the whales. And there was a couple of attempts to pull up the whales on this first-year ice and it didn't succeed. The ice was just breaking up. We ended up towing them to that ice pan."

Oliver Leavitt took longer to convince than some others that the climate had warmed. He kept hoping that the difference lay in the way people were perceiving the weather or that the changes were part of a cycle that would finally swing back to normal. But if it was a cycle, it was such a long one that no one could remember conditions like these, so bad for so long, nor did the elders tell of periodic warming and weakening of the ice to this degree. Evidence from tree rings and permafrost temperature measurements also indicated the warming was unprecedented and helped push Oliver over the line. Once he accepted that something was happening, many observations he had been holding in abeyance slipped quickly into the pattern.

"We used to get a lot more pressure ridges and heavy ice," he said. "Now it breaks away a lot more.

"Even the tundra thaws a lot more than it used to. Now, if you're making an ice cellar, you can go down four feet [before hitting frozen ground]. You were lucky to go down a couple of feet when I was a kid."

The shore seemed to be eroding faster. There was not much land left in front of Oliver's house. Big waves had a chance to build up in the absence of ice, and that could be contributing to the problem.

But Oliver wouldn't dwell on the subject. "Technology changes, the face of the earth changes, you change. Look at the Italians. They still eat pasta."

On May 7, the day after we scrammed from Oliver's camp, crews went back out on the ice. Then, in the evening, a ferocious current that looked like a river began flowing south to north. One moment the water was relatively still, a few minutes later small bergs were speeding by and water was flowing out from under the ice—I saw a large jellyfish emerge and shoot off into the distance in the clear water. This kind of current could break off the weak ice. The crews bugged out again.

On May 8, we returned to the ice. The sun blazed, surrounded by sun dogs, and the temperature was too warm for parkas, up to 34 degrees Fahrenheit. The snow was melting and water stood in puddles in dips all over the sea ice. It was unnerving to run a snowmachine through water sitting on top of ice that was sitting on top of 130 feet of water. Oliver had kept a haunch of caribou meat frozen in snow that was now disappearing. It was a traditional hunter's food: kill a caribou in cold weather, let it freeze, then carry the frozen haunch and snack on it, cutting chunks out with a sharp knife whenever energy flags or you feel cold. The raw, frozen meat was delicious: as it thawed in the mouth it released a bloody flavor like rare beef. Warmth soon returned to your cheeks. Techniques like these, using the Arctic cold to preserve food, helped the Iñupiat survive through long times of shortage and over journeys of great distance. But in this warm weather, Oliver's caribou had thawed. It was spoiled. He was seeing few whales. He said, "They've probably gone by. We're not seeing them in numbers. Probably gone by. Terrible year."

As the evening wore on, Billy Jens checked the crack behind us. He prepared to pack up for a quick escape. Oliver and the boys told stories of being surprised by whales, of accidents when someone mishandled the weapons, of close calls on the ice. At 1:00 a.m., the entire ice sheet we were sitting on dropped a little with a jolt. Soon the camp was packed again and we were retreating back down the trail with the sleds bouncing, crashing, and splashing over pressure ridges and through the slush and expanding pools.

The next day was warm again. The water was bright and motionless. The ice pack had receded dozens of miles from the shorefast ice. Many whalers on our part of the lead had given up. According to rules agreed to by the Barrow Whaling Captains Association, only the traditional, quiet style of whaling with the umiaq was allowed on the portion of the lead south of the tip of Point Barrow, or Nuvuk, which is about ten miles northeast of the town. North of there, whalers were permitted to use motorized aluminum boats—boats around fourteen feet long with 35-horsepower outboards, about as large as could be fit on a sled and dragged to the ice edge. The motor allowed whalers to range far afield to find a whale, but the noise chased the whales away from the shore. They called it "boating," and they called the boats "the aluminum." Some whalers thought their luck was so poor because those boating north of Nuvuk were diverting the whales offshore. In any event, with so few whales near the ice, there was a movement to allow the aluminum all along the lead. The decision was up to the three members of the executive committee of the Whaling Captains Association, of whom Oliver was one. The other two already had agreed. Oliver was among the very few still camped along the lead with an umiaq. He said he'd rather quit and go hunting upriver than whale with the aluminum.

"I'm out here whaling," he said. "To me, that's not whaling."

At 5:15 p.m., a prayer of thanksgiving came over the VHF, the harpooner of George Ahmaogak's crew thanking God for a safe and successful hunt—they had killed a whale. In accordance with tradition, the prayer not only announced the kill but alerted everyone in town to come help pull the animal up and butcher it. After the prayer, voices came on the radio with congratulations and cries of "Hey-hey-hey," a whaling cheer that sounded like the catch phrase used by Bill Cosby's Fat Albert on Saturday-morning cartoons thirty years ago. The prayer of Ahmaogak Crew's harpooner came through the little speaker on Oliver's VHF with a tone as thin as wrinkled paper in the still, damp air at the ice edge. It concluded, "In Jesus' name. Hey-hey-hey," and then a cheer came up from their boat, so many miles away across the water.

The Ahmaogak boat gave coordinates from their Global Positioning System receiver and called for more fuel and more boats to pull the whale. They were 14 miles north-northeast of their camp, which was, in turn, two and a half miles off Nuvuk. Their boat, appropriate for trout fishing on an inland lake, was attached to a whale four times its length and four hundred times its weight. Billy Jens, Brian, and Jens would go up and help with the butchering and claim a share of the whale for Oliver's crew, but there was no rush. It would take all night to tow in the whale.

Clouds blanketed the sky as night fell. At 10:00 p.m. it began to rain. The crew put a tarp over Oliver's seating area on the sled. This weather felt more like a wet spring day in Anchorage than whaling season in the Arctic. Oliver was disgusted. He recalled as a young man wearing two pairs of snow pants for spring whaling, standing night watch in temperatures 20 degrees below zero, trying to warm up by the white gas stove during the brief minutes when he was sent to make tea for the older men.

"Here's your global warming," he said. "It never rains this time of year. It melts the snow real fast."

The weather station in Barrow read forecasts over the VHF during whaling season so crews could ask questions at the end. On May 10, not long after the rain, the meteorologist announced in a tone of disbelief, "There is officially no snow on the ground." A foot-deep snow pack had disappeared in three days. Since 1940, Barrow's snowmelt had come ever earlier on an accelerating line. That year's May 10 was forty days earlier than snowmelt in 1940. Adjusting for the human-caused changes around the weather station (road dust in town enhances snowmelt) and using statistical analysis with a high confidence level, the snowmelt date had gotten eight days earlier, moving from about June 18 to June 10. Snowmelt on May 10 was off the charts.

As always, after the question-and-answer session on the VHF, someone said, "Thank you, weatherman." And the weatherman said, "You're welcome."

The ride was slushy and wet up north to the Ahmaogak Camp early the next morning. A parade of boats pulling the whale arrived at the ice edge at about 6:00 a.m., twelve hours after the kill. With little flat ice to work with, crews had made a boat ramp by smashing through a pressure ridge ten feet tall and fifty feet wide, one chop at a time, and many crews were using it to launch and retrieve their aluminum boats. Now it would be a whale ramp. Members of Ahmaogak Crew attached a heavy strap to the whale's tail. The flukes, a special delicacy, had been removed earlier to reduce drag while towing. A hand-held, gas-powered ice auger drilled a V-shaped channel in the ice several feet deep and about six inches in diameter, then crew members passed a heavy line through it. This was the deadman, or anchor point, for a block and tackle system. A set of two wooden and metal blocks each the size of a man's head had three wraps of yellow and black braided poly rope an inch and a half thick; the free end of that rope was attached to another, slightly lighter block and tackle, also with three wraps. Attached to the line leading off that block were the hands of about two hundred people ready to pull.

No one can own a whale. The whale gives itself to the entire community. The whaling captain bears the considerable expense of year-round preparation for the hunt and carries the weight of command. The crew provides skill and muscle and endures cold, danger, and tedium. Other crews come to help kill a struck whale and tow it back to camp. All the crews, and the entire community, converge on the ice to pull the whale out of the water and butcher it. Everyone who helps receives a share of the whale, as do elders and the infirm in town, and relatives far away, who receive care packages through the mail. The choicest cuts go to the captain and the crews who made the largest contribution. The captain's family then sets to work to prepare its share as a feast for the community, serving a banquet for all comers at the captain's house as soon as possible. In addition, for every whale caught in the spring, the successful captain's family serves the community at Nalukataq, an outdoor festival in June with the Eskimo blanket toss, a game in which jumpers are hurled high in the air on a walrus skin blanket held around the edge by many hands. Falltime whales are served at other special meals—each family has its favorite whale recipes for Thanksgiving and Christmas. In the end, the captain's compensation is intangible. He receives food for his family like everyone else, and the satisfaction that he has fed his people, and the honor and respect of the community—a fundamental kind of respect whose value does not fluctuate in the market of everyday relationships. These are not rules of whaling that can be broken or altered, such as those governing when a crew can use the aluminum or what day the hunt begins. You can't be Iñupiaq and own a whale.

At the cry of "All hands!" we stood together holding the yellow rope, a great, joyous crowd in the early morning, strung out several hundred feet along the ice trail leading away from the whale's tail, and someone cried out, "Walk away!" We walked, stretching the line, pulling harder as it grew taut, reaching the end of the trail and jogging back to the starting point to grab hold and keep pulling. We pulled hard, we got stuck, we were encouraged by loud calls of "Walk away, walk away," we moved again. Some strained to pull, some paced themselves. Some strong young guys were caught smoking behind their tents and put back to work. The whale made little perceptible movement. We'd stretched that long rope out, however, pulling the two blocks of the secondary, helper block and tackle together so we could go no farther. Now it was time to reattach the helper block and tackle to the heavier block and tackle set—to bring the helper line in so we could pull it out once again.

A dozen men threw themselves to their knees along the larger block and tackle and, weaving their fingers among ropes as taut as steel bars, twisted the lines together to keep the whale from slipping back while the helper rope was released and retied. The line slipped a little, the men twisted harder, a member of Ahmaogak Crew with a long-handled tool jammed rope in the block to stop it. The helper line was retied. A whale this size weighs well over 100,000 pounds, more than a fully loaded tractor trailer. In 1992 a ring holding the helper block separated—that was the same old whale in which Craig George found the stone spear point—and the thirty-five-pound block shot through the air like a cannonball, faster than the eye could see, hitting two women pulling the helper line. One was killed instantly, the other died soon after. A sad whale, as people still say. But the work goes on. The process of resetting the helper line must be repeated many times to pull in one whale. Two hundred workers picked up the line and the cry went up—"Walk away"—and again the line stretched and creaked with growing tension.

The day was hot, with broken overcast. Workers pulling the line threw down their coats and worked in flannel shirts and long underwear tops. A morning without breakfast was wearing on, and we were getting stuck more often. Fatigue had set in. Just in time, women from Ahmaogak Crew appeared from their tent, walking down the line with a big plastic bag of fresh homemade doughnuts and a metal cauldron filled with big chunks of boiled maktak, blubbery pieces cut from the skin of this very whale. Coffee and pop flowed freely, too. The maktak was best, tender and rich, a little salty, a little fishy, a little meaty, spreading a film of fat around the mouth and down the throat—really, like no other food. Tired workers gobbled fist-sized chunks and pulled again as if restoked with rocket fuel.

A flock of hundreds of eider ducks flew low overhead, bound toward the sea, their wings clapping and their voices ringing in thousand-part harmony. We were startled, and then the Iñupiaq people on the ice let out a spontaneous answering roar. A welcoming cheer, a cheer of pure exuberance for a bright day, for plenty, and for the prospect of fresh duck soup soon (although there's no duck hunting while whaling is in progress). The Iñupiat believe—sincerely, not just metaphorically—that each animal makes a gift of itself to the hunter to sustain the people. Today's abundance was like Christmas morning.

At noon, after six hours of continuous pulling, the whale lay entirely on the ice. The carcass was so large it was difficult to see. At great enough distance to take in the entire length, the details disappeared. Close up, the black skin rose like a wall to twice the height of a man. It hardly seemed like a living creature at all. Until I touched her. Then I could feel that this had been an air-breathing mammal, like me. She was still warm. Boys lay on top, basking in the warmth. The skin, as black and thick as a neoprene wet suit, felt smooth and yielding but strong, slick but not slimy, not at all like the skin of a fish. A team from Wildlife, including some visiting biologists, measured the whale: fifty-four feet, eleven inches. She was probably fifty to one hundred years old. I looked at a big eye, closed to a slit as if on the edge of sleep, and thought of all the ocean it had seen, under the Arctic ice and far out in the warm Pacific, over the course of most of a century. Eskimos were taking stock of the whale's size, too, their voices low and reverent. I felt humbled and grateful for this magnificent animal whose flesh I already had eaten.

The work of butchering began immediately. The whale had been dead too long. Bowhead spoil fast. The foot-thick blubber insulates them so completely that they don't cool after death. Wiry Brenton Rexford leaped upon the whale and began cutting with a blade on a long wooden pole, scribing rectangular slabs of maktak about eighteen inches wide and five feet long while a gang pulled the upper edge of the slab with hooks attached to ropes. As they pulled and Brenton cut, the maktak peeled back. When it was free, three or four of the hook holders ran away with the hundred-pound slab dragging on the ice behind them and swung it onto a sled.

Sleds quickly filled and snowmachines pulled them away. The ice turned red with blood. It poured out of the whale, flowing in little streams to crimson puddles inches deep. With the outer layer of the whale peeled back, workers began severing each vertebra, chopping with heavy tools while a dozen people heaved each piece free. They cut the black meat away from the bone, working forward one vertebra at a time, reaching the ribs, which looked like six-foot-long spare ribs. Someone tried a brand new chain saw, spraying blood around as in a slasher movie, but the whale got the better of it and the saw broke down. Others worked on the head, cut at the gelatinous sixteen-foot-long tongue, pulled apart the baleen, which was still entangled with the whale's last meal of shrimp, and noshed on the crumbly, pale gums from between the baleen strips, a bland, chewy snack. Only the bones and a few organs were not used—in the old days, the bones would have been used to hold up the roofs of sod houses. Some workers, such as Brenton, kept up a hard pace as the day passed by and evening came; more wandered away to their tents for a meal or a pop or sat joking with elders on the sidelines before setting back to work.

Victoria Woshner, a visiting Ph.D. veterinarian, seemed to relish pawing through the prodigious bloody entrails—the Iñupiat's inedible leavings—looking for scientific samples. Tall and slender in canvas coveralls, she waded up to her waist into a pile of foul-smelling intestines as thick as vacuum cleaner hoses, plunging her arms in here and there in search of the colon. I joined a group of men helping her extract the uterus. While we listened respectfully, Victoria lectured on the whale's reproductive history—it had given birth many times—and laid out the parts like an enormous diagram from a high school sex education class: ovaries the size of volleyballs, fallopian tubes like drier vents, and the uterus itself, as big as a queen-sized water bed. The group broke up and some of us returned to our epic struggle with the tongue. One of the men said, under his breath, "Man, that's one big pussy."

By 10:00 p.m. many people were too tired to keep working effectively. Some whalers were back at sea. With meltwater on the ice as deep as a foot in places, word had come that Oliver had quit whaling and dropped his opposition to using the aluminum all along the lead. Billy Jens and other guys his age with strength and authority were out boating. At the Ahmaogak whale, tired elders and middle-aged Eskimos sat around watching and admonishing teens and young men to keep at it; the youths would work briefly, then wander off again. Experts skilled in the jobs that kept the whole process advancing were few. A pod of beluga whales surfaced just off the ice edge and all the young people left their work to climb a pressure ridge where they could see white backs slipping along the surface in yellow evening light. A moment arrived when only one man was left working—one of the experts—and, looking around and seeing he was alone, he threw down his tool in disgust. By now the whale looked like it had been blown to bits by a bomb, messier than the carcass of a carved-up turkey. Hardly a recognizable piece remained except the skull, which held one eye high aloft on a tall tower of bone. A leader of Ahmaogak Crew jumped up onto the bones of the mouth and, in a booming voice, gave a short speech in Iñupiaq: he allocated the remaining baleen and tongue to the crews who had helped pull in the whale, and he called for the division of what had already been butchered.

Each crew had entered its name on a roll that recorded who was working on the whale. There would be forty shares plus the captain's share. Workers already had started rows of piles on a big, flat pan of ice back from the edge. Everyone gathered with snowmachines and sleds in an enormous circle while, in the middle, an old man and an old woman walked back and forth making sure the piles were equal and allocating the meat that was still arriving. A small team of young workers dragged heavy chunks back and forth to carry out their commands. I saw Brian Ahkiviana, the shy seventh grader from Oliver Leavitt Crew, still working hard, stripped down to a University of Alaska sweatshirt to stay cool, trying gamely to comply with the seemingly arbitrary and niggling orders he received to move a chunk from here to over there, then to take another, smaller chunk back that way, and so on. After days of little sleep and twenty straight hours of hard labor, his face was drained and his eyes were heavy. The piles each contained as much food as might come from a side of beef. Later, the crews would divide them further among their families.

Ahmaogak Crew struck camp, sleds of gear and whale meat speeding off down the trail one by one. Their captain, George, the mayor, had been busy at a borough assembly meeting when the whale was struck, but the honor of the hunt still belonged to him. (He would be up for reelection in the fall.) With his wife, Maggie, who was executive director of the Alaska Eskimo Whaling Commission, George needed quickly to serve the community and stash his share in his ice cellar so he could get on a plane to Japan for a meeting of the IWC that had already started. He sped by the crowd, sitting back grandly on his snowmachine and smoking a cigar, which he held aloft to cry out, "Hey-hey-hey!"

In the failing light, now past midnight and into the next day, the gathered whalers let out a cheer so loud it seemed to fill the big, deep blue sky.

CHAPTER TWO

▪

The Iñupiat

First Lieutenant P. Henry Ray, U.S. Army Signal Corps, and his ten-man scientific expedition arrived at Point Barrow on September 8, 1881, in a snowy northeast gale amid fast-forming sea ice. In the few days before the winter closed in completely, Ray needed a site to build an Arctic observatory and there to unload supplies from the small schooner, the *Golden Fleece*, that had brought him fifty-two days from San Francisco. Although the passage on the crowded ship had been unpleasant, it's hard to imagine the men were eager to disembark, either, as they squinted through blowing snow at an unprotected gravel beach barely showing above the breaking waves. Point Barrow was a logical place for Americans to begin studying the Arctic, the farthest north tip of the unexplored territory they bought from the Russians in 1867. But in heavy weather the point did not resemble the prow of a great continent. Instead it was a narrow, barren strand of gravel subsiding submissively into the waves at the end of the world.

Ray carried many scientists' hopes for the dawn of a new polar science. The Point Barrow station would serve as part of a system of stations taking identical measurements in a circle around the Arctic for the First International Polar Year. Previous Arctic science had involved expeditions to fill in blanks on the map and push a little farther north than the last group had gone—essentially, gentlemanly adventuring for personal and national tons of supplies there, struggling in waves that swamped his wooden boats and spray that froze to any surface it touched. With so few workers, the job seemed hopeless, and for days Ray cursed and raged at his men. The crisis turned when they got help and better-suited equipment. "The Natives, who at first appeared bewildered at the idea of our coming to stay, showed every disposition to be friendly now, and rendered us valuable assistance with their large skin boats (umiaks), and also in carrying stores up from the beach," Ray wrote in his report.

Thus began the cooperation of the Iñupiat and science. But it took Ray some time to recognize how valuable the local people really could be. First he turned his attention to building the station, setting up the instruments, and putting his men to work taking the temperature, humidity, cloud cover, wind direction, solar radiation, and various other quantities. After petty thefts by some of the villagers, Ray tried to hold the Natives at arm's length. Initially, it seems, he was uncomfortable with them. At the edge of the world, within the walls of the tiny station, Ray never forgot rank and discipline. The Iñupiat had none of that—there was no chief in their village, no system of social class, even families rearranged at will—and Ray seems at first to have regarded them as chaotic savages. But once the sheer tedium of the measurements bore down, he began to spend more time in the village. The expedition storekeeper, Ned Herendeen, a Yankee whaler who knew a little Iñupiaq, went along as an interpreter. Together they set out to learn about the unfamiliar place and people around them while leaving the rest of the party to read the instruments.

This arrangement intensely frustrated the expedition's most thoughtful scientist, John Murdoch. A naturalist from the Smithsonian Institution, he had been appointed a sergeant so he could go along on the trip. In his letters, Murdoch complained of being a lonely intellectual on the long expedition. No one else could carry on an interesting conversation, and Ray ranged in Murdoch's view from two-dimensional to despotic. But the sacrifices were all worthwhile to get to such an exotic place, meet the people there, and collect samples of their material culture and of the natural world. Then, on arrival, Ray ordered Murdoch to stay behind tending the instruments while he and Herendeen met the Natives, explored the area, and even attended an all-night Iñupiaq ceremony. Murdoch begged them to at least take detailed notes at the gathering, but when they returned to the station Herendeen's jottings were entirely useless. Moreover, Ray inglory. An Austro-Hungarian scientist and explorer, Lieutenant Weyprecht, weary of such dilettantism, instigated creation of an inter tional commission for a rigorous program of measurements of meteo ogy, earth magnetism, and the aurora. The polar regions were a gr scientific blank space—the other planets are more familiar now than t Arctic and Antarctic were then—and out of all the ignorance, Weyprecht three topics represented the most critical and attainable knowledge gaps t fill. With enough stations around the globe making identical measurements for a full year, he hoped patterns would emerge in the development of global weather, variations in the earth's magnetic field, and the shape and activity of the aurora. The program eventually enlisted eleven nations to mount fifteen expeditions, twelve in the Arctic and three in the Antarctic. Learned men attended a conference in Hamburg and other meetings to devise a detailed science plan, just as they do when big science launches an international project today. Then, like astronauts lifting off into space, teams on ships from many ports around the world voyaged into a hostile unknown, an icy void in the public imagination.

Ray lacked scientific training, but as a tough, self-made military man, he knew how to follow orders. His seven thousand words of official instructions went into fine detail on how to take the measurements—even how many notebooks to carry—and directed the men to read another forty technical books on the subject. The team would make multiple readings every hour for three years and every five minutes during two twenty-four-hour periods each month. The assumption implicit in the instructions was that a spot on a map could be known completely by sending a sufficiently diligent and well-supplied reader of instruments there. But when Ray arrived, it wasn't so simple. At first, he couldn't even find a site to build the station.

He needed to find a place quickly so the schooner could leave ahead of the forming ice. Inland, everything was swamp, and the shoreline itself was low and exposed to waves and ice movements. A few higher spots were already taken by the village of Nuvuk, at the tip of the point, and by Utqiagvik, about ten miles southwest, the 800-year-old village now called Barrow. Except for the Nuvuk village site, the point was a four-mile-long barrier beach of fine gravel periodically washed over by the sea. Ray chose a spot near Utqiagvik, on ancient village ruins (the area has been occupied for thousands of years), and set about unloading more than a hundred sisted on making all decisions about what artifacts to acquire. His eye went to the showy and fine—he wasn't interested in anything worn or utilitarian, the items Murdoch knew would be most important. Murdoch eventually broke away from the station enough to pursue his studies and, in the end, produced a detailed ethnological study that became a standard reference for the Iñupiat as well as for anthropologists.

The expedition's own final report concentrated on the numbers, the 480 pages of tiny figures the organizers had asked for. But it also contained an essay Ray wrote about his experiences. Although far shorter and certainly less professional than Murdoch's long treatment of objects and details, Ray's essay is human and full of charm, and a far better read. His naïve shock at Iñupiaq ways is evident—especially the role of women, who enjoyed far greater equality than in his own culture—but so is Ray's dawning respect for the Eskimos:

> Many of the old conservative men still cling to the habits of their fathers, and believe that stone arrow and lance heads possess virtues that makes [*sic*] them superior to those made of iron. They still teach the young men the art of chipping flint, and over their work tell them of the happy days before the white men came to drive away the whales and walrus, and when food was always plenty. An old man, when asked what he would do without the things the white men brought them, answered it would be very hard, and then to show us what he could do he showed a pair of boots he had on, and told us with great pride how, when his boots gave out while hunting, he killed a deer, made a needle from a piece of his bone, thread from the sinew, and made himself a new pair of boots from the skin, and asked, Could a white man do that?

Ray relied on these skills for inland forays by dog sled with Native guides, finding various sites never explored by whites, including the Meade River, which he named. The first statement in his report to Congress—the finding he seems to have thought most important—is "that the work of exploration in the Arctic can be carried on, at any season of the year, with the assistance of the natives, with comparative safety and but very little suffering." Elsewhere he noted, "the nearer one conforms to the habits of the natives the less liable he is to meet with disaster." By the end of his two-year stay, Ray worked and socialized with the Iñupiat; much of what he and

Murdoch reported about natural history the Natives told them. Ray even came to admire Native social practices: the Eskimos' solicitous care of disabled elders, their gentleness with their children, and the freedom of their spirit, in the joy and humor with which they met life.

Ray hoped that by cooperating with local Eskimos other travelers would find that the Arctic really wasn't so bad and the public would improve its opinion of the Far North. Instead, the opposite public impression developed. Ray's sister American expedition in the International Polar Year got all the press, and it wasn't good. Lieutenant Adolphus Greely led that party to the north of Ellesmere Island, at the extreme northern tip of Canada, and completed his measurements, like Ray. But with the third winter approaching and no vessel showing up with supplies or a ride home, Greely obeyed prearranged orders to attempt a retreat southward in open boats. The party never made it south of Ellesmere and spent the winter stranded on a hellishly barren island with volumes of measurements but too little food other than lichen, tiny shrimp, and the soles of their boots. Only six out of twenty-five survived. On arriving home, those six were accused of cannibalizing their comrades, who had died of starvation and scurvy, and one by execution for stealing food. The six denied the charge of cannibalism. Perhaps whoever cut incriminating chunks from the bodies died himself; nothing was ever proven. Greely's men might have survived if he, like Ray, had obtained geographic knowledge from the region's Native people. With a little information, he might have been able to reach the Inuit just across Nares Strait, in Greenland, who were marine mammal hunters like Alaska's Iñupiat. His loss was Ray's as well; the disruption of contact with Greely may have convinced the War Department to recall Ray's expedition after only two years.

Ray left behind the station and a big hole. Of all the expeditions' tedious and disagreeable tasks, among the worst was the order to dig a pit to measure the temperature of the frozen ground. Ray assigned the enlisted men. Other expeditions in the program dug a meter or two to take the measurements, but Ray wouldn't let up. According to local legend, he wanted to find the bottom of the permafrost. The diggers' log, covering fourteen months of intermittent work, is a laconic portrait of their frustration. The first five days of digging yielded a pit only five feet deep and five identical log notations of "Tenacious and very hard." Then, "Tenacious and very hard. Put in blast, which blew out without moving any earth." The next day, "Work suspended." But they started again, making as little headway as a few inches a day, until they had dug one of the biggest holes ever seen at Barrow. We have no record of what the Natives thought of this activity, but, practically enough, an Iñupiaq family appropriated the hole as a prodigious ice cellar after the expedition's departure and it was used for more than a century, like many natural freezers that preserve and age whale meat. The diggers gave up at a depth of thirty-seven feet, six inches, without reaching the bottom of the permafrost. They might as well have been digging to China: the permafrost at Barrow is more than a thousand feet deep.

The scientific legacy of the International Polar Year was valuable, but not in the ways the organizers expected. The voluminous reports were in many languages, making them difficult to use. The meteorological measurements were too far apart to explain much; scientists still debate today the underlying drivers of polar weather, which do indeed influence weather patterns over much of the world. On the other hand, the magnetic instruments for the first time clearly correlated the aurora and magnetic disturbances and captured magnetic storms among the largest seen for the next century on a widespread observing network, since the expeditions happened to coincide with a high sunspot cycle. That information formed the foundation of later understanding of disturbances in the earth's magnetic field.

The Arctic became an important setting for international scientific cooperation. The Second International Polar Year in 1932–33 brought scientists back to Barrow, and the International Geophysical Year, in 1957–58, which became a symbol of the scientific optimism of that era, was cochaired by a scientist in Alaska. The huge International Biological Program of 1968–74 included more than one hundred scientists in Barrow. Now another International Polar Year was in the offing, for 2007, to focus largely on Arctic climate change. And finally researchers were making good use of the meteorological data brought back by Greely, Ray, and the other expeditions 120 years ago. Thanks to the careful plan originally set by the scientists in Hamburg, the twelve Arctic stations were well distributed and the data well documented. For researchers trying to tease out long-term climate trends in the Arctic, this old data was just the thing.

Today, scientists were measuring Barrow more extensively than any other Arctic research site in the world. Researchers flowed through constantly to study climate change in ice, sea, atmosphere, snow, tundra, and permafrost—you could hardly turn around without bumping into a science project. A married couple of anthropologists from Philadelphia, Glenn Sheehan and Anne Jensen, adopted Ray's historic role of overseeing the scientific facilities and coordinating with the local population. Glenn directed the Barrow Arctic Science Consortium, known as BASC, a locally controlled science support organization funded by the National Science Foundation. Anne was the chief scientist for Ukpeagvik Iñupiat Corporation, or UIC, the scientists' Eskimo landlord.

Glenn first arrived on the North Slope to learn about the Iñupiat in the early 1980s, after leaving the navy, working on a doctorate in anthropology with a full scholarship from Bryn Mawr. His major dig was on Point Franklin, a long spit that reaches into the Arctic Ocean near the village of Wainwright, west of Barrow, where an ancient whaling village called Pingusugruk was disappearing into the sea. Glenn and Anne met at Bryn Mawr. Each summer, funded on grants, they traveled to a windswept camp, working and living in the cold and taking their water from a brackish tundra pond. Back in the Philadelphia area during the off-season, they made their living from contract archeology, investigating the sites of construction projects. In either place, the method that seemed to work best was to start by talking. Even in Pennsylvania the people on the land knew it best. The team found artifacts by canvassing the neighborhood before digging (although, working on set-fee contracts from construction companies, they made less money by finding more). In Wainwright, Glenn and Anne sought the elders' blessing and advice before digging in ancient village ruins.

Archeologists traditionally behaved much differently. Alaska Natives knew them as takers of objects, knowledge, and dignity, their practices epitomized by the grave robbers who fanned out to Native cemeteries from the world's great museums a century ago. Researchers from that period put living Alaska Native human beings on display as objects. Today, law requires museums to repatriate Native remains and artifacts, but a younger generation still has its own examples of unethical research. Harry Brower, Jr., told me about researchers who came to town asking about old bones, but without identifying their true purposes. He spoke openly of what he knew, and the next time he drove his snowmachine to his cabin, all the old landmark bones that marked the route for generations were gone. Consequently, researchers often encountered suspicion and reticence among North Slope people, and even from entire villages. One heavily studied community went to the extreme of claiming that all its cultural information, from the past or present, was its intellectual property. Before researchers could talk to elders, they were asked to sign a form giving the village control over anything they produced.

Glenn involved Wainwright in the dig because it was the right thing to do, and also because it made sense for his research. He expected some answers about the ancient people he was studying to come from above the ground, among those still living. Culture, he believed, was continuous. The ways people lived today could speak to how they lived long ago. Likewise, from the evidence underground, those still living might see the antecedents of their subsistence practices, their relations to one another, and even how they thought about the world—the roots of their own minds and hearts. The elders agreed the project would comport with the high honor they accorded their ancestors. Day by day, villagers came to the site of the dig to work and to talk. Elders knew the purposes and names of many of the ancient artifacts that emerged. As many as sixty villagers came at a time to help dig and interpret artifacts. Anthropologists and elders learned as equals.

Glenn studied Pingusugruk in part to build on the story of how climate change shaped the Iñupiat, which he told me as vividly as if he had been there to see it. Around A.D. 800, the Northern Hemisphere (and possibly the whole globe) started to warm up. The sun may have burned a little brighter; a natural 1,500-year cycle of cold and warm may have been reaching its peak. Glaciers shrank in Europe and North America, sea ice melted and drew back from shore, snow stayed higher on the mountains, warmth-loving plants grew farther north, fish were abundant in waters that would later be too cold. Winemakers thrived in England. The Medieval Warm Period lasted for four hundred years, long enough to help shape human history. Norsemen colonized Iceland and Greenland and even set up camp at the mouth of the St. Lawrence River. Their traders plied ice-free northern waters with exotic cargo. Europeans reaped good harvests, built Gothic cathedrals, and launched crusades to the Holy Land.

Cultural changes in the Arctic were even more dramatic. Before the warming there was a village at the base of the Point Barrow spit, a place called Pigniq. Ducks still fly over that spot in great numbers, above dozens of plywood summer cabins built for hunting them (its other name is Shooting Station). Artifacts from Pigniq dating from before the Medieval Warm Period represent an Arctic cultural tradition scientists named the Birnirk, a bastardization of Pigniq. Birnirk people hunted and traveled on the sea ice but were not serious whalers—to catch a whale through a lead far from shore and return with the meat was probably too hard. After the weather warmed, the sea ice became less stable, unsafe for Birmirk roaming after seals and other marine mammals, but whaling became easier in big stretches of open water near shore. A single family could do it, with perhaps one umiaq and supporting kayaks paddling in pursuit of a whale. The chances of success in open-water whaling weren't great on any given day, but just one whale could support a family for a long time.

The warming made the entire north a rich, open frontier for development based on whaling. Around 1000, the Thule whaling tradition (named for a dig in Greenland and pronounced TOOL-ee) quickly spread from Alaska all the way east across the north of Canada to Greenland, where Thule people ultimately ran into the Norse and learned to make barrels with wooden staves and baleen hoops. With the ice open all along the shore and whales abundant and widespread, Thule people could live anywhere they chose in small family groups without joining in large villages. Glenn imagined family conflicts arising and being suppressed by the culture's natural reticence, with the aggrieved party taking off to live somewhere else. As they spread, the Thule displaced or assimilated earlier Canadian peoples, wiping out a 2,000-year-long tradition known by archeologists as the Dorset (similarly, the Norman Conquest was going on at the same time in England). There is no question the expansion happened fast. Poor-quality Alaskan pottery not durable enough to last more than a generation turned up in Greenland archeological digs: it could only have gotten there intact if carried across North America relatively quickly. Alaska's Iñupiat and Greenland's Inuit share many legends and can understand each other's language, although they sound strange to each other today: a Greenlander told one Barrow Iñupiaq, "Hey, you talk like our grandparents used to talk."

Around 1200 to 1300, however, the climate changed once again. Perhaps the solar cycle cooled, or more volcanic eruptions shadowed the earth with ash, or both. The Little Ice Age began. In Europe, a six-year famine hit in 1315. The Black Death came in 1348. In the fifteenth century Britons gave up making wine. European agriculture eventually had to become more efficient and fishing fleets more robust, exploiting stocks far across the ocean. Glaciers advanced. Farmers in Iceland quit growing barley until the early twentieth century. Around 1350, the Norse abandoned their now-frigid Greenland colony. At the coldest, in the period from 1680 to 1730, ice surrounded Iceland most of the year, and one winter southwestern England and northern France had belts of sea ice. On several occasions, Inuit paddling kayaks showed up off the Orkney Islands of northern Scotland and once in Aberdeen, presumably having floated across the ocean on sea ice that broke off from shore.

The Thule way of life could not continue unchanged in the new, colder climate regime. The sea ice thickened and extended south and near-shore leads disappeared. In southern Greenland, the Inugsuk moved south into former Norse-controlled fjords and became seafaring people, living in big, communal houses. The Inuit of northern Greenland kept their Alaskan-style houses but forgot where they came from, thinking of themselves as the only people in the world. On the Arctic Ocean coast of central Canada, Thule descendants had to give up permanent dwellings, living during winter in igloos on the pack ice to hunt seal and during summer in tundra camps to pursue fish and game.

For the people of Alaska's North Slope, the new coming of ice meant life was no longer possible in the way they knew it. Climate change must have happened gradually. The ice froze thicker, lasted longer, and broke into usable leads farther and farther from shore. As the cold deepened, years came more frequently when whaling was impossible for people dispersed along the coastline, where most lived. At that point, Glenn said, the entire society had a decision to make. If survival demanded change, what should stay the same? What part of the culture was indispensable? As an anthropologist, he believed cultures would conserve their traditional food before all else. The Iñupiat kept eating whale.

To save the whale hunt, people had to congregate in a few places along the coast where the shape of the land created reliable open leads relatively near shore. The villages of Utqiagvik (Barrow), Nuvuk (Point Barrow), and Pingusugruk (Point Franklin) grew in these places, as well as a few others along the long coast, most of which are still occupied by Iñupiaq whalers. Even at these points, however, whaling was more difficult than it had been: whalers had to drag their boats miles across the ice to the lead and drag tons of whale meat back to the village. They needed more equipment, more workers, more coordination and preparation. People had to learn how to live in big villages with many interdependent families in close proximity to one another. In their digs and interviews, Glenn and Anne sought the evidence of these changes and, consequently, how the Iñupiat became who they are.

Instead of whaling as separate families, larger crews affiliated, each led by a skilled captain, a leader called an *umialik*. Like whaling captains today, the umialik was a man whose skills and leadership ability were recognized by his crew and who possessed the wealth necessary to outfit their hunt. His directions were unquestioned in the boat or amid hazards, but he wasn't an autocrat like the captain of a ship. The village lacked a formal hierarchy or any coercive authority. Everyone who joined with an umialik had an equal say in decisions, including women, each free to weigh the opinions of those with greater wisdom or experience. Not a system of voting or majority rule but one of consensus, in which each member was always free to go another direction. Respect held the community together and merit determined its leadership. Perhaps the arrangement emerged as a reasonable compromise for people accustomed to the nearly total freedom and autonomy of their previous life, dispersed wherever they pleased along the coast.

On this profoundly democratic basis, seaside villages grew by around 1600 to 300 to 600 residents living in permanent sod houses, with supporting populations in interior areas, previously sparsely populated, numbering as many as 1,300. In the new climate, whalers had to stay at the shore and concentrate on whaling and catching other marine mammals. Inland Iñupiat, without fixed settlements, followed the caribou to provide skins and meat for the whalers, receiving seal oil and whale meat in return—the caribou hunt was too unpredictable to sustain them consistently and they needed oil for heating and lighting and to dip their meat for flavor and fat. Family and tradition linked the groups, too, with joint celebrations, children adopted back and forth, and the loan of crewmen for seasonal whale hunts. Villages also traded up and down the coast, meeting in annual celebrations to exchange goods. The economics of comparative advantage meant each group was doing what it could do most effectively and relying on trade for other needs. But beyond simple trade, the coastal people also used their more reliable food supplies to carry their interior allies through the lean times so the system could be sustained.

Communication was critical. Besides individual family houses, the villages had communal buildings for gatherings and celebrations, each whaling group building its own *qargi* of whalebone. Villages had more than one and sometimes several, each under an umialik. Each group also maintained special grounds for its Nalukataq, the June festival with the blanket toss. Men congregated every day in the qargi for meetings and leisure and to prepare for whaling, hunting, and war. Its corollary today is Barrow's Volunteer Search and Rescue, where the VHF keeps track of hunters and whaling crews amid games of pinochle and billiards. The language and styles of social interaction—the aversion to conflict—may also have evolved to deal with the new social complexity and need for consensus: Iñupiaq has many words for "meet."

Whalers did not cook or make shelter on the ice, instead receiving support from shore. The harpooner's first thrust attached two or three sealskin floats with walrus-hide ropes (today they use poly rope and an inflatable buoy). The floats marked the whale's location and tired it when it dove. As many as twenty floats from six to ten strikes by various umiaqs might be needed to exhaust the whale. Then the hunters disabled the animal with a cut to a critical tendon, climbed on its back, and killed it with a lance driven into a vulnerable spot behind the head. Umiaqs towed the whale to the shore ice. There a woman would give it a drink of precious fresh water through its blowhole to assist the whale's spirit to return to be harvested once again. The crews butchered the whale as it floated next to the ice, cutting the maktak while spinning the carcass in the water like a corn cob. When enough weight had been removed, they could haul the other remains onto the ice by primitive pulleys and sheer strength of numbers. Any unusable portions were returned to the sea to assist the whale's return. Sleds pulled by dogs and people hauled everything else back to the village, where the bones became supports in buildings and the meat was frozen in ice cellars.

The system produced surpluses of food. Population grew and competition for resources developed. All good building sites were taken, as Lieutenant Ray found when he arrived. That led to war, which significantly reduced the population in the eighteenth century. Glenn and Anne's dig at

Point Franklin showed that warfare caused the demise of Pingusugruk. As difficult as the field work was, it fed Glenn with evidence for his theories about the Iñupiat and built the couple's relationship with the communities where they worked.

When I met Glenn and Anne, in 2001, they were well established in Barrow. Anne grew a garden of tundra flowers in the cold ground on the south side of their house, NARL building 170, a converted military Quonset hut of corrugated metal with numbers a foot tall still affixed to the outside. The growing season lasted two months and the gravel pad received half the annual precipitation of Tucson, but she persisted, even experimenting with a buried bathtub to retain the scant moisture. Glenn said his eyesight wasn't strong enough to see the blossoms without getting on his hands and knees. Anne smiled an inward smile, eyes down, to admit the garden didn't look like much. At first I took that smile as a sign of embarrassment, but after seeing it many more times I decided instead that she was smiling toward that part of herself that persisted in nurturing a tiny plot of Arctic ground.

Glenn's personality seemed the opposite of Anne's private nurturing. His voice was often dismissive or simply disgusted with what was going on around him. He cut himself the fewest breaks of anyone, analyzing his quick temper with the same mental agility and disinterest with which he dissected scientific puzzles. Physically small and wiry, he took on the world like a fight-crazed terrier. A smart but poor kid, he had joined the Naval ROTC for the education benefits, became an officer, and ended up running a big, unruly brig in Philadelphia, facing down dangerous criminals. He was generous and loyal and certainly invigorating to be around, but it seemed incredible to me that he had thrived among the conflict-averse Iñupiat. They hated to contradict anyone—it was one reason their meetings took so long.

But he did thrive. By the spring of 1988, Glenn felt at home in Barrow, although he couldn't afford to live there; he was writing a life history of one of Barrow's most respected elders, Tom Brower. Anne was pregnant. In the off-season the couple had rewarding contract work, studying the proposed route of a pipeline along the Schuylkill River west of Philadelphia. They had rented a house in the steep woods of the river valley suburb of Conshohocken with a workshop in a farmhouse a few doors down and had three employees: two young female archeologists and a burly ex-convict hired at the recommendation of the Salvation Army—Glenn was still involved in prison reform after his experience in the navy.

Coming back from breakfast on Good Friday, April 1, 1988, Anne went to the farmhouse to see if anyone had questions while Glenn made a phone call back at the house. As she passed the ex-con, Arthur Faulkner, on the spiral staircase to the second floor, he grabbed her from behind and threatened her with a four-inch hunting knife. He said, "You know what I want." Upstairs all was blood and overturned furniture. Clarise Dorner was dying there, raped and tied to a chair. Anne heard the other archeologist, Annaliese Killoran, known as Lisa, coming back from a coffee run and yelled for her to get away and get help. Arthur stabbed Anne four times in the back and once in the chest, then dropped her to the floor and ran after Lisa. Blood pouring out with each breath, Anne ran down the stairs and out of the house, down the steep hill toward the river, passing the patio, where Lisa lay in a fetal position with Arthur standing over her and stabbing her—there Lisa died. Arthur pursued Anne down the hill. She dove over a ten-foot-high retaining wall topped with pointed black stones and fell to an abandoned road where a pair of surveyors happened to be working. Arthur Faulkner lagged behind, looking for a gate. Anne told the surveyors she was pregnant and being attacked by a madman with a knife; they got in their pickup truck and drove away, leaving her.

Glenn heard Anne screaming and heard her screams cut off. He dropped the phone and ran to the farmhouse, found Lisa on the patio, heard her body making strange noises, tried to comfort her. He was attacked. The killer outweighed Glenn by 50 percent. He pinned him with one arm and stabbed him ten times with the other while they danced around Lisa's body. When Arthur reached down to take Lisa's car keys, Glenn broke away and ran. He knocked on a neighbor's door, then burst in, bloody and panicked. The man inside froze and did nothing; he refused to move.

Arthur Faulkner was caught later in New York City, tried, and sentenced to death. It turned out his prior offense had been for trying to rape a seventy-two-year-old woman who had befriended him through church. The families of the dead women sued Glenn and Anne and they, in turn, sued the Salvation Army, which had referred Arthur as a nonviolent offender. Through their long hospital stays and interminable legal proceedings their archeological contracts continued; instead of giving up and declaring bankruptcy, Glenn hired others to do the work, losing money all the time. They couldn't afford to move and lived several years more on Woodmont Road, the dreary street where it all happened.

When Justine was born, her parents knew right away that she had been injured, too. She couldn't nurse. Part of her face was immobile. It was years before she could smile. Each of the joyous changes that come upon children as naturally as leaves opening on a tree—sitting up, standing, walking, running—came late to Justine, with struggle, after predictions of failure. The diagnosis was cerebral palsy, but not the more common, degenerative kind. Doctors could never say confidently what Justine might be able to do. Loaded with unfair burdens, she inherited her father's rage and obstinacy. She communicated well in sign language, but when she learned to vocalize, she refused to sign any longer—although most people couldn't understand her speech. Nor would she repeat herself. But she made progress, too. When I met the family, Justine was fourteen and had recently made the compromise of saying things again or trying different words when she was not understood. She had marked the significant accomplishment of attending a movie with other teenagers without anything going wrong. She had developed an enthusiasm for *The Crocodile Hunter* show on the Animal Planet channel and for Jane Fonda exercise videos. On the other hand, when her school aide suddenly left town, she spent a week of class hiding in a padded box.

Glenn's toughness and willingness to fight helped them through. In the malevolent universe in which his family suddenly found themselves, lawyers seemed to view them with contempt and failed to defend their interests, school officials offered Justine the services she needed only after all-out war, and bills arrived even though the full-time job of taking care of Justine and all the other problems brought in no money. Everyone acted as if it was all Glenn's fault; sometimes they said so. He began to blame himself, too. The legal process ground on and on, ending only in 2002, when Arthur Faulkner's death sentence was commuted to life in prison. Even then, fourteen years later, Glenn and Anne remained deeply in debt.

Woodmont Road is a cul-de-sac, a vestigial stub of a road cut off by a passing expressway and left to decay. I went there. It wasn't easy to find, just a single, disintegrating lane under a dark tunnel of tall, spare trees. The entrance was marked by two signs: ONE WAY, DO NOT ENTER, and DEAD END, NO EXIT. I parked the car and walked. Highway noise roared from the other side of a narrow strip of tired, dusty forest, the Schuylkill Expressway, a congested section of interstate leading into Philadelphia along the rocky walls of the river valley. Two men emerged from strange little row houses set on the hill below the road to ask what I was doing there. No one walks down that road. They knew about the murders and warned me away from the white-painted stone farmhouse at the end of the road, still vacant after all these years. They said it was unsafe to go near. Vines and creepers had swallowed the lower walls and thick brambles blocked the walks and patio. With its varied roofline and the stone walls that angled into the hillside, it was probably a quaint country house once, but now time was tearing it down, rotting the wood, giving the dark glass a menacing cast. The house where Glenn and Anne lived had burned to the ground.

They escaped long ago, back to the excavation at Point Franklin and the labs in Wainwright and Barrow where they invited villagers to learn about their ancestors. Most people there, even friends, knew nothing about what had happened in Pennsylvania. Barrow's immense sky, cold, dry air, and unfenced tundra seemed in a different world than the musty dead leaves under the mournful trees of Woodmont Road.

In August 1994, Glenn and Anne had just finished a summer season at the Point Franklin dig when a storm eroded the shoreline of ancient Utqiagvik and exposed a human head wrapped in a parka hood, part of a frozen body. The declining sea ice cover had hastened erosion on this shore, where fifty prehistoric house mounds still remained. In 1982, an entire frozen prehistoric family appeared where they had been crushed in their home by an ivu. That discovery became a scientific and media circus and left the community and Glenn himself with bad memories of buzzing helicopters, morbid curiosity seekers, and disrespectful treatment of the dead. A hurricane-force storm came in 1986, when the ice was as far out to sea as many people could remember, and tore away more of Utqiagvik, including the rest of the frozen family's house and the burial of a man, possibly a shaman, who washed away before his grave site could be excavated.

Barrow elders didn't want either of those outcomes repeated, so they asked Glenn and Anne to develop a culturally acceptable research program for the new body, allowing just twenty-four hours to put the plan together so the situation would have no time to get out of control. Anne helped work out the ground rules with the community: the human remains could be examined, but only briefly, and then would be reburied. The body could not leave the state and photographs were not allowed. After study, artifacts would return to the Iñupiat Heritage Center. Glenn got on the phone, frantically trying to raise money for the work, knowing grant applications for science projects usually take months or years, not hours. The National Science Foundation agreed to help, but not right away. He needed $50,000 from Bryn Mawr to get started. "I'm telling this to the operator, and she's saying, 'There's nobody here. It's summer.' "

But he did get the money and set up a study like the one at Point Franklin, involving the entire Native community. Working together, they dug out a solid wedge of clear ice that contained the body—probably an ice cellar that flooded—then melted it free with warm water from a borough fire truck. They sifted the fascinating contents of the grave. Children picked up beads and elders identified traditional objects. It was only well into the work that Glenn realized the body was a child's. "The first thing we saw was the foot, and it was exactly Justine's size. It looked exactly like Justine's foot." At the time, Justine was six years old.

Anne flew to Anchorage with the body still frozen and met with a renowned pathologist, Michael Zimmerman, whom she and Glenn knew from the University of Pennsylvania. In donated hospital facilities she and the pathologist thawed the ice and unwrapped the duck-skin parka, feathers facing in, and the other clothing in which the little girl had been buried, eight hundred years earlier. The skin, flesh, and internal organs were perfectly preserved: not only was she the first ancient child to be found on the North Slope in more than skeletal remains, she was also one of the best-preserved bodies of any found. Zimmerman had what he needed to do detailed work while the body returned to Barrow for burial. She was named Agnaiyaaq, or Little Girl.

The autopsy showed Little Girl died with a severe disability, a rare genetic lung disease. A cruel myth about the Eskimos said they abandoned the aged and disabled on the ice to die (in the mass media, that image remained a socially acceptable subject of cartoons and jokes, which struck me as unimaginably racist once I was in Barrow). Perhaps it did happen in isolated cases during the social dissolution, starvation, and epidemics that wracked the Iñupiat after white contact. But when Lieutenant Ray was in Barrow, during a year of famine when the village caught no whales, he was impressed by how the Iñupiat paid aged and disabled elders utmost respect, even at great sacrifice. Around A.D. 1200, when the little girl died, her people were starving. The Little Ice Age was beginning, the ice was thickening, and great cultural change was under way to adapt to the great environmental change. Life was unimaginably hard. Hunting with stone tools and heating with seal oil lamps, people's lives rarely lasted forty years. But still this community had nurtured a disabled child like Justine.

"I was very impressed. In a way, I was very happy," Glenn said. "Here was a little girl who was disabled all her life and clearly was not going to be an economic contributor to the community, and she was taken care of all her life. She was buried with a baleen toboggan. Now, as archeologists, we have to make up stories about what we find. The one we imagine is that that toboggan was used to haul her around the village.

"What the little girl shows is that the disabled were honored, cared for and respected by the community."

Not much later, Glenn and Anne got their jobs at BASC and UIC, the focal points of Barrow science, leaving Pennsylvania behind for good. Their lives continued to revolve around Justine. It was easier in Barrow. The schools were excellent. The community accepted her with her disability. Scientists and Eskimo friends were used to her affectionate hugs, her unsteady gait, and her tendency to make loud, indistinct comments at inappropriate times. "If she falls over, or if she suddenly talks too loudly, they know her and they understand that," Glenn said.

He had a theory for why the community accepted Justine so easily. True to his work, the theory pointed to an environmental cause. Traditionally, villagers cooped up in dark sod iglus would sometimes go nuts in the winter and burst out raving, ripping their clothes off and flailing around. Everyone knew what that was all about, so they would ignore the outbursts without further mention. He theorized that Justine's outbursts fell in the same category.

I had a theory of my own. Iñupiaq culture, and perhaps survival, too, depended on community and cooperation—that's why the Iñupiat avoided conflict—and the binding force of community was mutual respect. You could have a disability, you could be an alcoholic, you could even go to jail, and the community still made a place for you. A North Slope Borough mayor involved in a massive corruption scandal returned to prominence as the fire chief and president of the Barrow Whaling Captains Association. In my own culture, which I didn't even know the name of (was it Western culture? American culture? dominant culture?), we existed as individuals. We valued people according to individual characteristics: ability, knowledge, wealth, physical appearance, and so on. Being part of a community was a choice—you could join the business community, or the online community, or a neighborhood community—but membership was always conditional. We liked to say respect was something you had to earn. For the Iñupiat, respect belonged to everyone.

Cognitive psychology measured this difference between individualistic societies and more communal Eastern societies (although not Eskimos, as far as I know). People in Western society tended to believe individual effort determined the outcome of events. Children learned that if you tried hard enough, you could accomplish anything. Consequently, we attributed bad events to the shortcomings of those involved—the way people blamed Glenn for what Arthur Faulkner did—and not to external causes the individual could not control. On the other hand, when viewing our own circumstances, we almost always found outside causes. This familiar bias was called the attribution error. Scientists found it in every society they studied, but it was much weaker in communal societies. To put it another way, those societies' mutual respect, measured and quantified, was greater than ours.

The people of Nuvuk defended their village in warlike fashion when the first Europeans arrived. Thomas Elson, an officer on HMS *Blossom*, had been sent by Sir John Barrow, second lord of the British Admiralty, as part of a mission to find the Northwest Passage. Elson commanded the ship's barge up the coast until he arrived at Point Barrow, which he wanted to name "World's End," at 2:00 a.m. on August 13, 1826. Nuvuk was the biggest village the expedition had seen along the coast. After an initial attempt to trade boat-to-boat went sour, twenty Eskimos fended off Elson's landing, brandishing spears and bows and arrows, and he turned around and returned south to the *Blossom*. The point ultimately was named for the boss back in England and the large inlet inside Point Barrow became Elson Lagoon.

To the west, Point Franklin was named for Sir John Franklin, a British officer exploring the coast to the east of the *Blossom* at the same time. Barrow sent Franklin back to the Arctic again in 1845 with two ships and 128 men to attempt the Northwest Passage from east to west. Franklin disappeared. Barrow's written orders were so vague no one knew quite where to look for the expedition, so the Admiralty dispatched many ships to search the passage, attacking the problem from both the Atlantic and the Pacific. A vessel named the HMS *Plover* had a small supporting role: it was a slow, ugly depot ship sent to resupply Franklin should he emerge on the west end of the Northwest Passage in Alaska and to support the ships sent by the Admiralty to search that part of the world. (The British ultimately spent twelve years and more than 750,000 pounds to find Franklin's bones and log books near where his ships had frozen in place midway between the two oceans.)

The *Plover*'s captain, Rochfort Maguire, stood this tedious duty station anchored two miles east of Nuvuk, waiting to be of assistance to Franklin's ghost. He had a good friend along from earlier service together, Dr. John Simpson, the ship's surgeon, who was as much of an expert on the Iñupiat as existed at the time, able to speak some of the language and driven by scientific curiosity to learn more about them. Captain Maguire was a remarkable man too: despite his military background, he was compassionate in his dealings with the Iñupiat, setting out from the start to make friends. His careful journal records a two-year voyage of discovery while standing still, as the British sailors and Iñupiat villagers overcame fear, developed mutually beneficial relationships, and then formed real friendships.

An *angatkuq*, or shaman, named Erk-sing-era became the visitors' closest friend and confidant, calling on the frozen-in ship every day or two during their stay. One can sense Maguire's wonder through his daily log entries as his respect for Erk-sing-era grows. Simpson joined Erk-sing-era on medical rounds in the village, helping tend to injuries, illness, and malnutrition while the two doctors noted each other's techniques with professional courtesy. After observing Simpson's skill, Erk-sing-era asked what fee would be appropriate for Simpson to produce a favorable wind to return a stranded umialik and companions who had drifted away on the ice. In conflicts instigated by either side—thefts by the Natives or, critically, the accidental shooting of a villager by a sailor—Erk-sing-era negotiated proper reparations and probably prevented war. The ship's crew helped the village, too, feeding starving families and commiserating with those who had lost members on the ice. They invited the Natives to join in their holiday pantomimes next to the ship (a scene I enjoy imagining). Maguire even gave Erk-sing-era powder and shot for his gun, the only Iñupiat gun in that region, a supreme measure of his faith in a potential enemy.

Like many scientists who followed, Dr. Simpson found the Iñupiat to be his most valuable source of knowledge. He relied on Erk-sing-era for discoveries in natural science as well as ethnography. The primary British interest was geography. Erk-sing-era drew a map of the coast east of Barrow. Simpson reported it was accurate in every detail the British knew about except it left out one mountain range they had charted; in fact, Erk-sing-era was right—mountains don't exist where the British map showed them. In his monograph on the expedition, Simpson showed the two maps superimposed to demonstrate Erk-sing-era's accuracy and to take advantage of his knowledge of huge areas of Alaska unexplored by whites.

In 1853, Erk-sing-era and his young son came to Christmas Eve dinner on the *Plover*, as Maguire recorded in his journal:

> He tasted almost every dish, each of which he pronounced better than the preceding one, but without the slightest display of gluttony or greediness such as is commonly attributed to men of his race, and which in the present state of scarcity of food at the village might have been expected. He seemed to get weary of the singing after two hours and went home at four o'clock, saying he was well pleased and what a glorious story it would make. . . . The child ate moderately too, as many a British child would not, but unfortunately got sick from smoking tobacco too freely. He soon recovered however and joined in the chorus of some of the songs the men have been teaching him during the winter.

When Lieutenant Ray arrived at Point Barrow thirty years later, some Nuvuk adults who had been children remembered fat, thick-necked "Magwa" and his men. Erk-sing-era was still alive, but feeble and bent over double, walking with two canes, quite deaf and nearly blind, although still sharp mentally. Ray and Murdoch got the impression white contact had not caused much change: "In all their intercourse with the whites they have learned very little English, chiefly a few oaths and exclamations like 'Get out of here' and the words of such songs as 'Little Brown Jug' and 'Shoo Fly,' curiously distorted," Murdoch wrote.

But they were wrong. The New England whaling fleet had discovered the bowhead whales of the Arctic coast in 1848, and by the time they reached as far as Barrow, in 1854, they had had killed more than 7,000 whales. When Ray arrived, the number killed exceeded 15,000. (By way of comparison, a century of recovery has brought the total population back to about 10,000.) Bowhead were scarce and the Iñupiat were starving. Sailors had introduced alcohol, prostitution, and diseases that ripped through a Native population without immunities. The coastal villages shrank and the interior region was entirely depopulated. When explorer Vilhjálmur Stefánsson surveyed the coast around 1907, he found few who had been born there: as coastal people died, surviving inland Iñupiat had migrated to fill the umiaqs and carry on the whale hunt. A few years later a Yankee whaler witnessed the death of the last Iñupiat interior village after its members caught the measles at a coastal trading celebration attended by sailors. Believing they would recover if they went home, the villagers perished one by one on a desperate trek upriver, the bodies of men, women, and children scattered along the riverbanks with the gifts they had received at the celebration and their empty boats.

Iñupiat society broke down, but whaling survived. Crews joined in new combinations, around families, an umiaq at a time. In the 1880s and 1890s, with whales scarce and fewer hunters surviving to pursue them, Eskimos adopted Yankee whaling techniques that could increase their success rate and reduce the manpower and coordination necessary. They set up wall tents and cooked out on the ice. They killed whales with brass pipe bombs fired from the harpoon shaft and from huge brass shoulder guns. They landed whales for butchering with block and tackle. The new tools also made the hunt safer: with enough bombs, you didn't have to climb on the whale's back to kill it.

Technology transfer went both ways, with Eskimo whalers offering as much know-how to Yankee whalers as they received. Eskimo harpoons had a head with a hinge, or toggle; the entire head would turn perpendicular in a wound, making it much harder to pull out than the Yankee whalers' simple barbed harpoon head. New England whalers adopted that innovation in 1848, probably from Greenland Eskimos but possibly from the Iñupiat; the timing suggests the idea wasn't entirely original. The Eskimos then adopted back the iron version of the toggling harpoon head brought from New England.

The Eskimos' shore-based whaling also had advantages over Yankee methods. Every year wooden whaling ships were "nipped," their hulls caught and crushed by ice. Losses increased in the 1870s as ships pushed ever farther into danger to increase their catch of depleted stocks. And sometimes they just had bad luck: in 1871 the entire bowhead fleet was caught in the closing pack ice just west of Point Franklin, destroying thirty-three out of forty ships (although the 1,219 men, women, and children aboard were rescued by the remaining seven ships, and no lives were lost).

Whaling from shore made more sense. That was the brainstorm that came to Ned Herendeen, the storekeeper of Ray's International Polar Year expedition during his two years at the Point Barrow research station. For a small fraction of the cost and risk of whaling from a ship, a boat crew could be stationed in Barrow over the winter, ready at the coming of spring and late into the fall to hunt whales where ships couldn't reach. When ice conditions threatened, the crew could wait on shore, like the Natives. In San Francisco after the science expedition, Herendeen found a backer for his idea in the Pacific Steam Whaling Company. The company cheaply rented the abandoned Signal Corps station from the government and dispatched Herendeen with twelve men and supplies. But in two years he caught no whales, while the Iñupiat villagers caught many. He was fired and ten new men were put in his place. George Leavitt led that crew, with Charles Brower as one of his boat steerers.

In their first year, Leavitt and Brower caught only one whale, while the Natives caught twenty-two. Unlike Herendeen, however, they were wise enough to see why they failed and the Iñupiat succeeded. Most important was the boat. The umiaq was perfect for hunting from the ice. Their own wooden boats were rigid and heavy. Moving them over miles of ice to the lead without damage was an ordeal requiring the men first to chip perfectly flat trails through the pressure ridges. The skin-covered umiaq was light, easy to move, and flexible enough to withstand extraordinary punishment. It slipped through the water fast and quietly. In 1888, Brower gathered an Iñupiat crew and, observing their ceremonial requirements, whaled successfully. Shore whaling in Barrow exploded for several years. That was when the Iñupiat, seeing the success of the newcomers' methods, adapted completely to their new, hybrid style, which they kept using after commercial whaling disappeared in 1907. A Paris fashion show in that year rendered valueless the last bowhead product in commercial use—baleen for high-fashion corsets—by introducing a new, more natural look, in the process saving the whales from extermination.

Charles Brower stayed in Barrow and raised a large Eskimo family. Several of the most important Iñupiat elders of the late twentieth century were his sons, including Tom Brower, Harry Sr., and Arnold Sr. In 2002, Arnold Sr. was the last of the sons living, still nominal captain of one of Barrow's most successful crews, the ABC, or Arnold Brower Crew, although his son, Arnold Jr., captained the boat now. Arnold Sr.'s children and grandchildren were more than experienced and numerous enough to consistently land whales. His eleven daughters (including Maggie Ahmaogak) made it unnecessary to hire helpers to stitch the umiaq with caribou sinew, a job that cost other captains around $2,000. Several other important families in Barrow bear the names of New England whalers, too, including Adams, Bodfish, and Hopson. Oliver Leavitt's grandfather was George Leavitt, Charles Brower's original boss at the Barrow whaling station (Oliver says George Leavitt also had families in Hawaii and New England).

After the fall 2001 whale hunt, Arnold Brower, Sr., and I talked for hours while he cut up black whale meat to ferment six weeks for Thanksgiving dinner. As was customary in traditional Native culture, he started the conversation by tracing his lineage—even the legend of a man from Nuvuk who carried Eskimo ways across Canada to Greenland.

"Our ancestors come from Nuvuk," Arnold said. "I was born into Yankee whaling."

He told of changes in the climate, told stories of adventure, such as rescuing a crew when a whale bit their umiaq in half, and even told how to catch a whale.

"I learned all this from the elder people," he said. "We know the whale movement and migration. The whales know their time. All animals know time. They know when to get out for survival, and when to come back for reproduction. We've got to know that for subsistence. When to collect them. We didn't learn that overnight. We're not perfect, I'll tell you that much. But we've learned enough to survive and make it."

The Iñupiat know their environment, and they know how to create and use technology—to find creative solutions to practical problems. That has been a key to their survival since the beginning of the Little Ice Age. Oliver

Leavitt told a story of elders stranded on the tundra with a broken-down snowmachine who got home by fabricating a replacement bearing from driftwood pieces hardened with oil and fire. He himself built in plywood, epoxy, and fiberglass. He could make a boat lighter and handier in the water than a traditional skin umiaq, one not easily punctured on a careening sled or needing expensive recovering every two years. Some disapproved when these methods were introduced. The old women needed the money they earned sewing the ugruk skin on boats. But catching the whale isn't a sport, and Oliver believed any technology that could make it safer and more successful would benefit the Iñupiat. So he reached into his imagination and pulled out a boat, working long days for several weeks in the Heritage Center, without plans or specifications, cutting wood by the feel and look of it, fitting pieces together as best they fit. A frame of hardwood, upside down, bent downward at each pointed end into graceful concentric curves, fastened with pegs, bolts, and screws, stiffened with thin plywood sheathing, swathed in white fiberglass, and covered, near the gunwale, with the advanced fabric used on inflatable boats. Oliver made mistakes—gaps here and there to fill with epoxy—with a curse and a cheerful "We're not building a watch." But the boat, as it finally emerged, was truly beautiful and handled just as well as he had hoped. As he said, "The boys like it."

In the room where Oliver worked, the Heritage Center posted core cultural values on posters and the backs of business cards: respect for elders, others, and nature; family kinship and roles; sharing; knowledge of language; cooperation; love and respect for one another; humor; hunting traditions; compassion; humility; avoidance of conflict; and spirituality. Before the arrival of the outside world there was no need to spell out those values, but even in the wired Barrow of today the list still sounded true. Like a precious liquid poured from cup to cup, the Iñupiat had passed a way of living together in the world forward through time, through epidemics, social breakdown, and climate change. They passed it on even as the cups changed that held this cultural essence. Oliver remained an umialik, whatever his boat was made of.

Glenn Sheehan came to see the culture as a necessary response to the need to consistently harvest the whale—for the sake of the food, and for all the meaning derived from it. The environment itself programmed the Iñupiat for pragmatism and adaptability. That belief gave Glenn faith they could handle today's environmental change, too.

"The way whaling is being carried out right now, certainly in the spring, can become more difficult and hazardous, and perhaps less productive," he said. "But we know that when it was warmer a thousand years ago, they were doing fine with whaling. And remember, it's not that people are attached to going three miles or eleven miles out to the lead and building a camp. That's not the thing. It's conducting whaling and coming back with whales and distributing the product to the community and working together throughout the year leading up to the whaling season and celebrating afterwards. That's what's important. And if the format of the whaling itself changes a little bit or changes a lot, we've already seen that it doesn't make any difference. And I can't imagine that it will make any difference now."

That sounded quite reasonable sitting in comfort in NARL Quonset hut 140, surrounded by Glenn and Anne's books, scientific journals, and copies of *The New Yorker*, listening to Glenn's thoughts flow out rapidly in an unreconstructed Philadelphia accent (saying "wuder" for "water"). Not many people knew more about surviving hard times. But climate change didn't seem so easy with Oliver, sitting in the rain on puddled ice, worrying about breaking off into the ocean, and watching him resist using the aluminum boats—the more practical way of whaling in bad ice with lots of open water.

Besides, science couldn't say if climate change would return us in a circle to the Medieval Warm Period or if we would go far beyond one millennium back and beyond the scope of human experience. Paleoclimatologists have learned to look into the deep past, reading the records left in corals, peat, ancient pollen and insects, lake levels, ice drill cores, plant fossils, cave deposits, tree rings, snow and mountain glacier levels, windblown dust, boreholes under the ocean, into the continents, and on glaciers, and other subtle indices. Sixty-five million years ago the earth was an average of 12 degrees C warmer than it is today (the current global average temperature is about 14 degrees C). There were crocodiles and palm trees at the poles. The climate remained stable, cooling very gradually, for many millions of years. Then, around three million years ago, the climate began swinging ferociously back and forth between warm and cold periods (the glacial periods popularly called ice ages) on cycles timed roughly with the earth's rotational wobbles and slightly eccentric path around the sun. It was as if, at that point, the earth system became far more sensitive to the sun's changes, like a musical instrument resonating at a sympathetic vibration. Three million years of seesaw cooling and heating may have given human beings the chance to evolve, favoring our adaptive ability to deal with sudden and extreme changes in environment. But it is not inevitable that the earth system will remain at that sensitive pivot point of warm and cold. Princeton's George Philander has theorized that the warming induced by human emission of greenhouse gases could tip the intricate, resonating system of ocean currents, snow, and ice back into the warm, stable state that existed before our species emerged. "The likelihood of pushing ourselves back to three million years ago is not that far-fetched," he said. "We're actually pretty close to it."

Even without such an extreme scenario, the Iñupiat may face changes their ancestors would not recognize. Climate changes of the current interglacial period, the non-ice-age era that spans all of human history, have been small enough that paleoclimatologists have a hard time teasing them out of the record. They probably amounted to 1 degree C, including the Medieval Warm Period and the Little Ice Age. In the twentieth century, temperatures are probably sweeping upward faster and more consistently than in the record of the previous millennium, and the warming has been especially pronounced in the Arctic. We may already be passing through the climate of the Medieval Warm Period on our way to something else.

CHAPTER THREE

■

The Snow

SIX SNOWMACHINES ZOOMED through the wide, winding corridor of a river canyon. The sky was brilliant blue and the vertical canyon walls, with billows of drifted snow, reflected the sun like a mirror, pure white. Yellow Ski-doo machines, swooping like swallows, interwove paths over smooth, virgin snow, crossing the flat river bottom and tracing big, graceful loops up the steep sides. Sparkling brightness rushed by, too quick to absorb, crisp air roaring in gulps too big to swallow. At each bend in the frozen river a new scene unfolded, as trackless as if unexplored, an irresistible invitation deeper into wilderness. Six riders flew across the Arctic snow on powerful machines, healthy, tanned, and utterly free.

This was the *Mission Impossible* team Matthew Sturm had assembled, each member picked, like those on the TV show, for a particular skill to contribute to the task of measuring snow all across Alaska's Arctic. Matthew's fellow principal investigator (or PI, in science jargon), Glen Liston, of Colorado State University, held degrees in atmospheric science, snow and permafrost geophysics, and applied mathematics: he knew how to re-create the real world of snow inside a computer. Jon Holmgren, a tall, gentle man with a bushy beard, was the gear head, an inventor and self-taught machinist from Fairbanks who knew how to make sensitive scientific equipment from scrap metal and fix anything that broke. Eric Pyne would have been the muscle man in the cast—a tough Fairbanks outdoorsman, owner of a small sawmill next door to Jon's machine shop, who relished long days in cold weather on a snowmachine with a burrito cooking on the hot engine under the cowling. Ken Tape was Matthew's right hand, his shaggy and enthusiastic young grad student (a role, in science, like an indentured apprentice), who perpetually looked as if he had just been woken from a sound sleep, his hair standing straight on end day after day. April Cheuvront was the only woman, just twenty-six years old and the only member of the crew who had never been on an Arctic expedition before—in fact, as a North Carolina middle school teacher, she had never encountered real winter before. Everyone was in high spirits, but April's smile was as broad and blissful as a newlywed's.

April, Eric, and Ken crested a sixty-five-foot-high river bluff while the three older team members stopped down below. Above the canyon the land around was utterly flat, crusted with wind-hardened snow just a few inches deep, but where the Meade River cut this corridor and disturbed the wind's flow the snow had gathered into a pillow-soft drift that clothed the cliff down to the river. April and Eric leaped off the edge, sliding on shovels or on their snowsuits, tumbled down, then crawled back up, swimming against small avalanches of their own making, and slid down again. Ken descended slowly, planting a long aluminum probe to gauge the depth of the drift. Down below, Matthew sighted on the probe with a theodolite, a surveying instrument like a computerized telescope, while Jon recorded the numbers he called out in a yellow field notebook. I stood by and watched, having arrived as a passenger on the back of one of the machines. The white canyon walls focused the sun's lively spring photons on my face; I could feel that radiant warmth even while I felt the convection of the day's cold air cooling the shadowed side of my body. Every shout fell as flat as a deflated beachball, lost in the soundproofing of the snow.

Glen looked around at wind-formed shapes and explained what he saw in a deep voice, nearly a whisper. He saw this landscape with a mathematician's eye. After more than a decade of working in the cold, he had developed a complex set of equations that, when run on a powerful desktop computer (a Sun workstation), could accurately re-create the snowscape around us. Given the precipitation, wind, and topography, Glen's computer model calculated the erosion and deposition of the snow on the landscape at each point on each day, using the strength of the snow surface, the level above the snow at which the wind would either pick up snow grains or simply bounce them along, the quantity of snow that would disappear into vapor (a process called sublimation), and the way the wind's flow would be affected by hills, dips, and bushes. Glen knew how the snowdrifts looked in the computer. He looked at real snowdrifts to make sure they matched. Not particular snowdrifts: he had spent years in the field developing a feel for snow, an intuition about how it behaves, so that he could recognize realism whenever it emerged from the equations.

Matthew's experience of this world was more tactile. He gloried in measuring snow. "I spent about a decade of my life doing nothing but measuring the R value, the insulation value, of snow," he said. That was one of the first things he told me when I first called him at his Fairbanks office of the Cold Regions Research and Engineering Lab (CRREL), a branch of the U.S. Army Corps of Engineers. Matthew said he was the world's leading expert on snow's insulating properties, a sufficiently narrow boast that I had no trouble believing it. Later a colleague told me Matthew was also the world's leading expert on seasonal snow in general. That was a bigger deal: all snow, covering a large part of the world for a large part of the year. Working with his team of men with icy beards, Matthew had crossed Alaska's coldest, windiest reaches many times, taking thousands of measurements, fascinated by variations in the hardness and thickness of snow in places where no one else ever bothered even to set foot. Matthew's internal energy seemed to keep him warm. Digging his fiftieth snow pit in a random spot on an endless, featureless plain in a biting wind, he could become as excited about the layers he found there as if he were uncovering the Rosetta stone. He seemed to know intellectually that others were not as fascinated by the contents of snow pits as he was, but deep down he couldn't really believe it. His drive worked like an independent force within him, a radioactive core that he sought to shield from the world but that flashed through nonetheless when he heard or saw something that really interested or surprised him.

Preparation for the journey had taken two years. The team planned to cross Arctic Alaska from Nome to Barrow in five weeks, 411 miles as the crow flies and more than 1,000 miles by the snowmachine odometer. The trek would cross over the tundra and through the scrubby trees of the Seward Peninsula, into the rocky Brooks Range, and across the exposed, flat whiteness of the North Slope. Matthew called it a transect, one long slice across Alaska to expose a cutaway profile of the snow.

Hardly anyone does trips like this. Eskimos rarely have reason to, and most scientists wouldn't know how. Most Arctic science in Alaska is a summer-only business. Lots of scientists who specialize in the area have never gone there during the cold season, and a surprising number have never seen the region they study at all. Caleb Pungowiyi, a Native elder and science leader in Kotzebue, told me, "Scientists are like tourists and geese in the springtime. They come marching up in the spring and go back in the fall, when some of the real changes they should be observing are in the winter, with the temperatures, the snowfall, the winds. They always tell us, 'We have to teach in the wintertime.'" Some simply don't like the cold. Others say automated equipment can tell them everything they need to know about winter.

But automation certainly wasn't the answer for snow. It couldn't even tell how much snow fell on the North Slope. In these windy conditions, snow tends to blow over the top of a snow gauge, skewing readings by a factor of two or three. The weather station in Barrow records "trace" snowfall around 190 days a year—meaning some snow fell, but not enough to measure accurately. The official record adds up all the trace readings to equal zero, but even a little bit of snow multiplied by 190 could be a significant amount. And no one was able to say exactly where the snow came from—did the same storms bring snowfall north and south of the Brooks Range?—or how much of it melted as opposed to sublimating away into vapor.

Matthew surely would be studying these questions no matter what, but institutional interest (including salary and logistical support from the NSF for the Nome to Barrow transect) derived from questions about climate change. Arctic snow helps regulate how much heat the earth absorbs from the sun. Snow reflects about 80 percent of the sun's warmth; snowless ground reflects only about 20 percent and absorbs 80 percent. Thanks to this reflective quality, called albedo (from the Latin for whiteness), the snow over the top of the world cools the planet like a foil dashboard shade on a car parked in the hot sun. If a warmer climate shortened the period when snow covered the ground, as appeared to be happening, the earth's albedo would drop and the planet would absorb more of the sun's warmth. That could cause further warming of the climate, which could, in turn, further shorten the winter season and reduce the albedo yet more, in a spiraling pattern known as a positive feedback loop.

Even if the term is new to you, a positive feedback loop has been demonstrated to you when your ears have been assaulted by a screeching microphone at a school assembly. It happens when the input of a system, in that case, the principal's microphone, is affected by the output, the gymnasium's loud speaker. The school PA system picks up its own noise and amplifies it, adding more noise with each cycle until it overloads with a loud, distorted screech. That's a positive feedback loop—one that leads toward instability. Another example: a dieter gets depressed, eats a lot, gets fatter, gets more depressed. A negative feedback loop is one that tends to dampen the system—the output reduces the input—such as a thermostat that turns off the furnace when a room is too hot and turns it on when it is too cold. Or a dieter who eats a lot, notices weight gain on the bathroom scale, curtails eating, and returns to normal weight. The climate contains many feedback loops. Thanks to the negative feedbacks, such as the cooling shadows of clouds billowing on a hot, sunny day, the climate is stable within a range that can sustain life.

Some of the strongest positive feedback loops are in the Arctic. Most of the world varies between green and brown on land, blue or green on the sea, absorbing roughly the same amount of energy from the sun year-round. Antarctica is white all year. But in the Arctic, the land and sea are white part of the year and dark part of the year and the extent of change in the albedo depends on the warmth of the climate. Theory and experience suggested that this positive feedback warmed the Arctic faster than other parts of the world, and that the accelerated warming could, in turn, drive more warming everywhere. The exaggerated rise in temperature was called the Arctic Amplification.

Sorting out climate feedback loops is one of the most important and difficult challenges of global change research, revealing enormous gaps in our understanding of how the earth works. One has the illusion, in our media-immersed society, that science knows almost everything, lacking just a few genes to unlock cancer, a final push to land on Mars, and a better computer chip to create artificial intelligence. But the frontiers of science do not spread in a smooth circle. The map of scientific understanding looks more like a spray of roads and pathways punching into a big, unexplored wilderness. These knowledge roads have been built up with discoveries, but it is strip development: if you peer behind the highway storefronts you see nothing but empty country out back. That is where the questions about Arctic physics and ecology lie. To put together the Arctic system, science needs to fill in vast gaps of unexplored real estate, figuratively and literally. Matthew estimated that a third of his transect was on marked trails, a third in places sometimes visited by locals, and a third where no one ever goes.

Matthew and Glen were used to such territory. They had worked for years on the North Slope and on the southern side of the Brooks Range, at Council, near Nome. Now they wanted to pull back to a broader scale. It seemed reasonable that the southern side of the Brooks could serve as a surrogate for how the northern side would look after warming. Computer models that predicted the climate's future were hard to build—perhaps impossible—but this might be a working model provided by nature. Or it might not: the mountains are high, and for a storm to affect both sides it would need to find a way around. If different storm systems affected the two sides, they would produce different snow and different climate profiles and comparison wouldn't mean much. So Matthew and Glen decided to take their team across the Brooks Range on a single transect to find out.

They would compare the snow two ways. Chemical snow sampling might be able to establish the ultimate source of the moisture—what ocean it evaporated from or, at least, if it evaporated all from the same ocean. Snow strata in pits dug each ten kilometers along the route could tell more. Matthew knew how to read each layer's story of snowfall and subsequent change by temperature, wind, and humidity rising up from moist soil. Lying on the snow, looking through his glass at crystals that looked like extraterrestrial gemstones, he could reconstruct the winter from a single pit. It was a craft, old-fashioned science based on expertise, careful observation, and elaborately detailed notes. Matthew relished each crystal. Even on the coldest days, he was happy lying down with his head in a snow pit.

Some people fear cold and never give it a real chance, like children who won't try a new food. Others, who enjoy winter sports, have at least some tolerance for it, an ability to bundle up, get active, and feel the regenerative thrill of thriving on a cold, snowy day. But extreme cold is never fun. Not many people have the ability, like Matthew Sturm and some Eskimos I met, to overcome it and behave as if it doesn't exist. Not to tough it out but simply not to care.

Matthew couldn't be sure how April would react to the cold, and that threw an unaccustomed element of the unknown into his planning. When he agreed to take a teacher as part of an NSF program called TEA (for Teachers Experiencing the Antarctic and Arctic), he insisted on the right to reject anyone about whom he had doubts. There were so many ways it could go wrong: a female teacher could be intimidated by being alone with the men, or could flirt and create jealousies, or could sit back and be a princess, or could develop a chip on her shoulder, or, worst of all, could simply shut down, unable to deal with the cold and remoteness. The problems wouldn't be physical, he said. "It's always spirit. You have to want to be where you are."

April had doubts, too. Afraid she might not be tough enough for the transect, she started lifting weights well in advance and running thirty-five miles a week. She had already been running twenty miles a week and had hiked a two-week backpacking trip from Great Smoky National Park to her home in western North Carolina, accompanied only by her dog. She enjoyed the effort and the quiet of backpacking. She felt happiest in the woods, alone or with family, simply enjoying the peace and beauty. The transect wouldn't be like that. It would be long days of hard work on lifeless snow and ice with five men she didn't know and no chance of escape. She had never been on a snowmachine. She had never experienced a northern winter. As she prepared herself and her classes for the trip—she wanted her students' year to revolve around it, too—she knew the journey would be entirely unlike anything she had ever experienced.

April's experience of the outdoors began when she was a child in West Virginia and went on trips in her family's camping trailer. Her father was a firefighter and her mother was a teacher. She met Steve Cheuvront as a senior in high school when they were both lifeguards at the local pool. He was four years older, but despite her father's initial attempts to protect her from him, they dated, backpacked together, and married. Economic opportunities were so few in West Virginia the couple moved far away to North Carolina, where Steve opened a one-man legal practice and April began teaching science at a big middle school down a hilly country road near Morganton. That was five years ago.

Despite her narrow background, April seemed a good bet to fit in on the Arctic transect, in part because she fit in so poorly in her regular job. In a school where teachers marshaled students to the lunch room in silent single-file lines like inmates in a prison, April's class was an oasis of independence and creativity. Her students collected water samples from a watershed threatened by a controversial lakeside development project, analyzed them chemically, then produced reports with graphs and spreadsheets on laptop computers. Sitting in the teachers' lounge, where she rarely went, she looked like an outsider, some kind of visiting academic, fit and tanned in her earth-tone knitted dresses and Birkenstocks among pasty women, mostly a decade or two older, and stereotypically southern. She listened politely and didn't mention the Arctic while they talked of journeys to Dollywood, the country music theme park. At dusk, April escaped up a long, twisting road, 4,000 feet high into the mountains. She and Steve lived in a little mountain cabin with knotty paneling and a big Franklin fireplace, surrounded on all sides by the massive, dark trees and cool, wet earth scent of Pisgah National Forest.

Matthew wanted to test April before the transect and planned two cold-weather trials. As soon as she arrived at his home in Fairbanks he took her to pick out a beaver pelt to be made into a hat by his favorite furrier. April normally would never wear fur, but she went through with it with a wary smile. On a second trip she flew to Nome with Matthew, Glen, and Jon. April learned to take snow measurements and ride a snowmachine. Matthew watched how she handled a chilly eighty-mile ride to Council, the team's research site, and the days of hard, cold, dark work, and the intimacy of a small, remote cabin. April had decided to prove her worth by always grabbing the coldest, nastiest job before anyone else had the chance. She rarely stopped smiling. Matthew was impressed. In the pictures from Council she looked tired and slightly disoriented, her head wrapped in brown fur, but cheerful, as if aware that this was all an extravagant practical joke. April never gave in to the cold, even on the worst days. But, unlike Matthew, she certainly felt the temperature and, as she said, the "hurtful" wind.

On March 20, 2002, the six snowmachines pulled out of Nome bound for Barrow. Each pulled sleds built in Fairbanks, some by Jon, of white, ultra-high-molecular-weight plastic, smaller but more durable than Eskimo sleds of lumber, with bowl-shaped sides and sturdy fabric coverings. Everything the team needed rode on the sleds in plastic Rubbermaid tubs called ActionPackers, lashed down by Glen's careful system. Stopping every ten kilometers to take measurements, Eric complained the machines never warmed up enough for their engines to properly cook anything. Each stop was part of Matthew's plan of more than 34,000 measurements in all, including the snow's depth, hardness, and water content, how the snow attenuated light, close examination of its layers, and ultra-clean chemical sampling.

At night they set up tents. The cooking tent was heated by a small wood stove fed with willows, which grow as low shrubs on the North Slope. The computers stayed warm in a heated portable office, an ingenious little room on a sled that Jon had devised, with solar power and a propane heater that kept it above 50 degrees F for the entire trip. A Honda generator recharged batteries nightly, including those for the Iridium satellite telephone. April used that link to post updates to her Web site—photos, journal entries, and lessons for her class back in North Carolina. Matthew knew he was driving his team hard, with long days of tedious measurements in wind and temperatures as low as –30 degrees F. He intended to keep them as comfortable as possible under the circumstances.

Matthew wanted no adventures and, as far as he was concerned, there were none. He chose each measurement or camping site in advance, deployed a remote cache of fuel and supplies eleven months early, and set up contacts with the schools in each village on the way. Matthew had done some mountain climbing as a young man, but the transect was work, not recreation, and an adventure meant a mistake, potential danger, and maybe a lost research opportunity. On the worst day of the trip, the wind started howling while the team was still working, visibility dropped to nothing in blowing snow, and some members of the team began to worry to themselves about the need for shelter and how they would set up the tents in such a storm. Matthew didn't stop working until conditions made work physically impossible. Then he called it an early day and prepared for the five-mile trip to that night's planned campsite. Without being able to see anything, he drove the lead snowmachine following a course on a handheld Global Positioning System receiver. He knew there was a river drainage that would offer shelter from the wind somewhere along the way. In the direct blast of the blizzard, a broken-down snowmachine or a team member who became separated from the group would mean serious trouble. When he saw a willow close by his machine, he knew he must be near the drainage. At the camping spot indicated on the GPS a riverbank provided enough shelter to put up the tents. Not long after, the team had eaten a hot dinner and picked a DVD to watch on the laptop—April got to choose *Shakespeare in Love* because they had already watched *The Matrix*. "We have notes right up to the point we realized the blizzard was getting dangerous," Matthew said. "I wish the taxpayer could see that."

After the Brooks Range, on the flat North Slope, there were two weeks of utterly empty country, without sight of another person and with hardly any animals. Just encountering snowmachine tracks was a noteworthy surprise. Communications were spotty or just too much trouble to set up in a tent camp after an exhausting day of work, so there was little word from home. But the team was working well; it had become self-contained, without much need of the rest of the world. April had proved herself as a hard worker, a creator of happy moods, and a fully integrated member of a group of friends. The chemistry was like that of adult siblings who have worked out their childhood rivalries—that unspoken intimacy, adjustment to familiar differences, and stability. Some of the guys bet on how long it would take April to decide to move to Alaska.

Then they arrived at Atqasuk, an Iñupiaq village of 230 next to the Meade River about sixty miles south of Barrow. There, for the first time since their departure, they saw cars, and April felt the jolt of reentry. At the BASC-operated house where the team stayed (and where I joined them) Eric, Ken, and Jon settled back on the couch after dinner to watch a war movie on cable TV. That made April uneasy, too. After months of worrying about whether she would measure up as an Arctic explorer, she was beginning to worry about returning to civilization. Out on the snow the world was simple and pure and relationships trusting and straightforward. She had come to think of that as real life and the messy indoor world as strange and uncomfortable. I've felt the sensation myself after stints in remote places, a prickly, antisocial feeling of being hemmed in by walls and human ugliness. April had felt it after returning from long backpacking trips in the Appalachian Mountains, but on this trip she was coming up from far deeper in the wilderness than she had ever been before. And Atqasuk was only the outmost fringe of the inhabited world.

The process of reentry to civilization was familiar, even effortless, for everyone else on the trip. Besides many previous Arctic expeditions, most of them lived in Fairbanks. It was not unusual for the temperature there to get down to –30 degrees F, and when it did, people still went on about their usual business. It was a sprawling little city, but snowmachines were legal within the city limits—they went speeding down the Chena River through the middle of downtown—and at the edges, past the last junk lot or suburban house (which could be one and the same), a rider could continue into hilly wilderness that extended thousands of miles. The porcupine-style resentment against crowded humanity that April began feeling in Atqasuk—many people in Fairbanks had that wilderness disease all the time.

After dinner and the intensely irrelevant movie, the evening dimmed. The team set out from the BASC house for the short walk to the Meade River School. The metal-sided building served kindergarten through high school and looked much like any other public school, except that it all stood on pilings, including the swimming pool wing. North Slope Borough oil money gave these villages excellent schools. The science teacher had offered the computer lab for the transect team to check personal e-mail. As the computers booted up, the mood came down. One by one, voices died as e-mail accounts opened and brought forth weeks of messages from home. The expedition's communications did not allow time to read e-mail at leisure, weep a little, and write long, heartfelt messages back. The session went on in silence, except for clicking keys, past midnight, until they walked back to the house, one by one, in the dark.

Matthew didn't believe in going out early, as that's when the world is coldest, and he liked his people to eat huge, fatty breakfasts. In Atqasuk a cook in a small construction camp served up big, cheesy omelets, sausages, and hash browns. Cholesterol was not a worry when you spent the day working hard in cold weather. (A Barrow biologist and musher who trekked to the North Pole reported that he lost weight and lowered his cholesterol on a diet of 7,800 calories a day, 60 percent of it fat, including a crunchy stick of frozen butter with each meal.) Breakfast would be the only leisure period until bedtime, and everyone knew that downing as much fat and as many calories as possible now would mean staying warm later. April dug in even though the food was not to her taste. At home she was a vegetarian, mostly. Her only meat was the venison her husband, Steve, brought home from the woods. Now she packed on calories and muscle and worried about the change in the shape of her body. The guys teased her about it every day.

The team suddenly was ready to go and sped off onto the snowy tundra, a precipitous way of leaving like jumping fast into cold water (the Eskimos did it the same way). Once off the trails in the village, the machines and sleds jolted viciously over ridges carved by the wind into the surface of the hard snow. Running in the same direction as the prevailing wind, the ride was smooth; cutting across the wind's direction, your spine was hammered repeatedly to the seat.

Matthew claimed he had been out here enough to find his way around, but it was hard to see how that was possible. It was like navigating at sea. There was texture but no shape. The sky was gray, the ground was white, the horizon was far away, but no telling how far, in an even circle all the way around. Native elders knew how to navigate by the wind-carved ridges in the snow. Matthew rode up on one knee, holding a small GPS unit in one hand, adjusting course right and left to zero in on a spot twenty-three kilometers from Atqasuk that he had chosen from a map when he and Ken set up the sampling system back in Fairbanks. When we arrived, the spot looked identical to everywhere else. Matthew leaped from his machine and threw a handful of snow aloft, watching how the wind caught it. Then he sent a couple of the guys out to find a lake shown on the map. They wandered around digging in the snow with aluminum shovels. Forty percent of this terrain is lake, but in the winter the lakes are invisible. You have to dig a hole to find one inches below your feet.

The team set up the site with the efficiency of a whaling crew. Jon measured out a hundred-meter line for detailed measurements and began digging pits along it. Glen and April departed for a nearby site in white clean suits and thin plastic gloves to gather snow samples for chemical analysis—this was one of the bad jobs April had volunteered for, freezing in uninsulated gloves for hours while depositing snow in plastic bags. Eric and Ken set up instruments to measure the light attenuation and hardness of the snow, one managing the computers inside the office sled while the other manipulated the devices outside. Matthew lay down with his head in one of the pits Jon had dug and marked the layers of snow, measured their thickness, inspected the shape and size of the grains through a lens, and weighed samples on a sensitive scale to determine their water content. I booked for him, writing down the numbers he dictated in a yellow field notebook, feeling my fingers chill in skiing gloves—I had taken off my thick, outer mittens to be able to write. Looking up from the page, I saw caribou walking on the horizon, gray and distant but distinct in shape, like cardboard cutouts mounted on the sharp rim of a big circle that surrounded us. Then I turned back to the notebook to catch up with Matthew.

The transect's 415 pits had already yielded the transect's first, tentative discovery. Matthew believed that precipitation here came from the same storms as on the south side of the Brooks Range.

Matthew learned this skill of snow stratigraphy, and much else—the enthusiasm, the ability to disregard cold, the focus on big issues—from Carl Benson, his Ph.D. adviser and the previous umialik of Alaska snow measurement. Matthew gave Carl credit every day; he talked about him with the reverence that Oliver Leavitt had for the elders who taught him to whale. When Carl called Atqasuk, Matthew took the phone with bright-eyed eagerness and afterward called Carl "a great man" and traced his lineage as a researcher to famous Arctic explorers who passed through Barrow, where Carl first worked in 1948. Both described their relationship as like father and son.

Carl had learned the study of snow layers from his own Ph.D. adviser and mentor, Bob Sharp of Caltech, who now was ninety-two years old but still in frequent contact, like an elderly parent. Carl dug snow pits on the Greenland ice sheet for four years to produce his own Ph.D. dissertation, which showed where and how the sheet acquired the snow that built it. Over four years, his team developed some of the skills Matthew was using, including the ability to keep working through bad weather. When Carl came to the Geophysical Institute at the University of Alaska in 1960, he recognized the unfilled niche of studying Alaskan snow. Two decades later, he recruited Matthew to be his student. Matthew, in return, got to do his master's thesis on a glacier sitting on top of a volcano in the Wrangell Mountains. "What you offer a student, a really outstanding one, is not money or the prospect of a special position but the opportunity to work on something they really want to work on," Carl said.

It was Carl who showed that the Barrow snow gauges were missing most of the snow and estimated how much. He also studied where the snow was going by measuring the drifts on the Meade River beginning in 1962—the same drifts Matthew's team would measure and play on forty years later. The river's canyon walls were steep, and high enough to catch all the snow blowing across in a winter; by measuring this snow trap in selected locations, Carl was able to calculate how much snow the wind carried in a winter. Matthew didn't have anything particular in mind by measuring the same drifts again, he just did it every year when he passed by Atqasuk, keeping a long-term data set alive in case it was ever needed again—and as a tribute to his mentor.

Ken Tape was supposed to be the next scientist in the line of succession from Sharp to Benson to Sturm. Matthew praised him fully, sounding much like Carl praising Matthew, and talked about the father-son, mentor-student bond they had developed. Ken had become skilled at snow stratigraphy and worked as hard as Matthew needed; he worked like a beast. Ken started coming to CRREL to help Matthew from West Valley High School in Fairbanks; five years later, working on a master's degree, he had already seen his name as a coauthor on a paper in the journal *Nature* that received world media attention. Ken was well aware of the opportunity offered to him. But the truth was, he didn't like the cold. He had learned how hard these trips could be on his first traverse, from the Brooks Range to Atqasuk, running exhausted at Matthew's pace, suffering in temperatures that never broke above –25 degrees F, constantly wishing he were somewhere else. Ken's sleeping bag wasn't warm enough, so he lay awake shivering many nights. He didn't know how to simply disregard the weather as Matthew, and Carl before him, could do. Day after day the flat whiteness, the cold, the wind, and the presence of the same few faces bore down on him. He swore he would never be back.

When I asked Ken about Matthew's ambition for him to carry on the tradition of snow science, he hedged. Maybe he could work in the Arctic in the summertime, he said.

Ken was a gentle and quietly sympathetic young man behind his perpetual bed hair, bleary eyes, and regular-guy manners. When anyone else had trouble with a snowmachine, Ken was the one who stayed back to help. He checked on the less experienced people—April and me—to make sure we were warm enough, had enough to eat, had our gear in order. He knew what it was like to suffer on these trips. The others weren't inconsiderate, just oblivious to the hardships, especially Matthew, who could keep working happily in a blinding ground blizzard without noticing the sandblasting snow. Ken admired and appreciated Matthew, wanted to hold up his end of the bargain in the relationship, took on the research program with real excitement—but it didn't look like he was going to become another Matthew or another Carl. But then, often a son doesn't see his future as his father does, and sometimes he ends up there anyway.

At the site outside Atqasuk, midafternoon arrived before I realized we wouldn't be stopping for lunch. I dug into a big bag of one-pound chocolate bars on one of the sleds. The others said they were sick of chocolate, but I was new to the expedition, and to me it seemed like the fulfillment of a childhood fantasy: all the big Hershey bars I could eat, without consequences. My folly became evident before the first bite. I couldn't break the chocolate. I rewrapped it and banged it on a snowmachine. The chocolate shattered. I put a piece in my mouth, but it had no taste. It felt like a piece of cold stone. Slowly, the warmth of my mouth melted the chocolate in oozing layers that I could suck and swallow as a sticky, viscous liquid. I put the rest of the pieces in my pocket and tried to hide my disappointment.

The team kept working without pause. When April and Glen returned from their day of chemical sampling, we split up. Glen and Matthew finished up an in-depth site while Ken, Jon, April, Eric, and I ran another twenty kilometers—almost halfway to Barrow—to start another site. It was evening; some yellow sun slipped through the crack between the ground and gray clouds. We flew across white open tundra, jolting on bumps in the hard snow that the flat light and speed made it difficult to see. When we arrived at the site, Ken jumped off his machine and threw a handful of snow in the air in perfect imitation of Matthew, and everyone laughed. The guys flopped on the snow for a snack and talked about what they would eat and how they would spend their first few days when they got home. Ken planned to drink beer and vegetate for a few days. Jon wanted to sit in the middle of his shop just to be surrounded by his tools and equipment. April didn't speak up. In a few days she would have to resurface into the world of people and cars, back to the foothills and mountains of North Carolina, from deepest winter to the flowers of late spring. Even Ken acknowledged that life is easier on a transect—not more comfortable, but easier, without complications.

I dug the pits, chopping at the snow to force the metal through it, and set up tall sculptures with the resulting snow bricks, just as Jon had shown me how to do that morning. This was a necessity so that we could imagine what someone would think in the improbable event that anyone ever passed this way. My sculptures looked vaguely Aztec, geometric sentries standing on the plain of white. Then the wind knocked the largest ones down.

Jon walked off into the distance poking the ground with a rod. He had invented a snow-depth probe with a magnetic collar that moved up and down; when he poked the rod through the snow the collar stopped at the snow surface. A push of a button then activated the device to log the depth of the snow—the distance between the tip of the rod and the collar—and the exact location, downloaded through a Global Positioning System receiver. Before this invention, snow researchers made depth measurements manually, poking a graduated probe into the snow then writing down the depth and estimated location in a field notebook. Jon's device, the Magnaprobe, allowed him to take precise measurements at an even walk. At each site, he walked away into the distance, taking a kilometer of measurements in a straight line, an insanely cold and tedious job if done by hand.

The Magnaprobe was a tool for a fundamental problem in understanding and predicting the climate, the problem of scale. The case in physics may be more familiar. At the scale of atoms, weird quantum laws describe a jumpy reality based on probabilities; at the scale of tennis balls, Newton's classic laws work just as well, and they're much simpler, easier to use, and more explanatory of the world we live in day to day. In the realm of climate, we are the size of atoms. Whether an umbrella or a parka is necessary on a single day for a single person is essentially meaningless in relation to understanding the entire climate system, yet that's the statistically jumpy scale at which life occurs. Starting from that small perspective, we want to understand the large-scale, Newtonian-style laws—the smooth, overarching principles that make it all comprehensible. More than that: we'd like not only to know the trajectory of the Newtonian tennis balls but also to predict how the tennis game is going to come out. The temptation to believe in our own omniscience is intense—just log on to the National Weather Service Web site and watch the storm outside your window from the perspective of a satellite above the earth. But omniscience is an illusion based on forgetting the many intermediate levels of complexity you can't see, abetted by the abstraction of modern, indoor life. Spending time outdoors, forced to accept your own small size, you perceive the world as progressively larger and more mysterious. Vulnerability teaches respect for the power and complexity of nature.

Climate science regarded Arctic snow two ways: as an undifferentiated white sheet and as molecules of water freezing and thawing. Without knowing if their assumption was correct, scientists used the particular to represent the general. Then Matthew Sturm, Glen Liston, and their snow research team found on ever-lengthening transects that the snow varied drastically in its thickness, insulative properties, response to wind, and other climate variables in a patchwork over many scales. It wasn't just Matthew's personality that drove the need for so many measurements over such large areas; that was the way, without shortcuts, to describe snow from the scale of individual grains to the entire region. Jon's Magnaprobe snow measurements over a kilometer, repeated scores of times, filled a gap between a small and a larger scale. Still, Matthew wasn't ready to say he knew anything about Arctic snow in general, just that he knew the snow he had touched, across the breadth of Arctic Alaska, just one slice of the world's Arctic.

When the team reunited, evening was well advanced, the sun hanging at a horizontal angle. They were a little giddy from the long day of work and the relief that it was almost over, at ease briefly in a laughably arbitrary spot on an enormous plain of white. The teasing and inside jokes that never stopped for long broke out in shouts and fists full of powdery snow that blew away before reaching their targets. Someone suggested a group picture. The team lined up in front of the office sled for a photo with long shadows in the yellow light. I shot a few with each person's camera. They stood quietly waiting. I could hear the wind. The mood dropped. Passing cameras back and forth, similar pictures in each camera bound for a different home, meant the trip was almost over, this might be the last normal day, and these same people likely would never be together again, certainly not this way. After five weeks, every other place, all other people, had come to seem unreal and far away. But soon, this time and pure white space would dissipate into a dreamlike memory instead.

We arrived back at the house in Atqasuk after 11:00 p.m., not unusually late. Some people prepared dinner while others began the work of data management, which took as long as three hours every night. Matthew, Glen, and Ken downloaded data loggers attached to the instruments into laptops and Zip disks, spot-checking each file to make sure the equipment was working properly. Jon and Eric attended to broken instruments and machines. Each day's work produced seas of numbers—the device that measured the strength of the snow produced thousands of measurements in a few seconds' run. But there was no time to figure out what the data meant. Its only use would be to produce answers to particular questions the scientists would publish in narrowly read journals, with luck to get noticed, cited by other scientists, and added to the canon of their small craft. After that, the raw data was likely bound for oblivion in an archive.

This process of deriving knowledge, so utterly contrary to the way the Iñupiat (or most anyone, for that matter) gather and use information to guide daily decisions, could make science seem irrelevant and even useless. Why freeze outdoors measuring snow for months just to publish a paper years later that only a few specialists will read? But the process yields answers that can be gained with certainty no other way. For example, an answer to the question posed by Carl Benson almost fifty years earlier, "How much snow falls on the North Slope?" Glen's computer model resolved it using Matthew's data. He ran the model many times with varying initial conditions—snowfall quantities—until the snow contours that emerged inside the computer matched what the team measured in the real world. The input that produced that accurate output had to be the real snowfall. At the same time, Glen discovered that 40 percent of the snow that fell on Alaska's windy Arctic coastal plain disappeared back into vapor through sublimation without a chance to melt and contribute water to the ecosystem.

While the scientists worked with their data, April worked on a laptop for hours every night, writing her on-line journal and class lessons for her students back home. Glen helped her post photos from the digital cameras and proofread her essays for the errors produced by late nights, with bleary eyes, after fourteen-hour work days in the cold. She never missed a day. With the four-hour time change, she could upload a unique Arctic science lesson for the teens back in Morganton each morning before bed, just in time for their first class at 7:40 a.m. Like children opening presents Christmas morning wrapped by their parents only hours before, April's students would never know the price she paid to produce those chirpy, information-packed messages for them each morning.

Matthew was worried about April's reentry into civilization. When you're on an expedition, living in a state of social deprivation, you think frequently about the people you've left behind and imagine they are thinking about you, perhaps planning a grand homecoming. But when you get home, you find their busy days have given them other things to think about and your place in their lives has filled in like water smoothing the surface of a pond. Matthew said returning from an expedition was a chance to see how little your world would change if you died. After some forty expeditions away from home over twenty-five years of his career, he had learned to avoid getting depressed or angry afterward, knowing in a week he would again become necessary to his wife, Betsy, and his children. But this would be April's first time. How would she react to finding herself erased?

Asking around the village for an elder knowledgeable about snow led me through Bernadine Itta, a secretary at the school, to her father, Thomas Itta, Sr., age seventy, whom I met when he came to the cafeteria to eat lunch with the children, as he did every day. He was thinking that day about making ptarmigan snares in the willows with the leg sinew of caribou, a skill that he learned at six or seven but that children weren't learning anymore, and made a plan with the principal to take a few students out to show them. Thomas himself had reached only the fourth grade, but he was a teacher with profound understanding of the area around Atqasuk and how to thrive there, deeply respected by both the white teachers and the Iñupiat. He had eight children and thirty grandchildren, most of them still in the village—a significant chunk of the population—and had lived there almost since its founding in the mid-1970s. His wife, Flossie, grew up in a camp on the Meade River near where Atqasuk came to be. Thomas hunted and ranged far afield on his snowmachine every day, despite his age.

After lunch we sat at his house near a table with a half-finished jigsaw puzzle while Flossie sorted pieces, watched television, and contributed sporadically to the talk, sometimes in English and sometimes in Iñupiaq. Thomas was lean and healthy-looking, sharp-eyed and brimming with information. Listening to him was like following the very smart and active scientists I was used to interviewing. As the master atop the food web, he was as wiry, fit, and aware as a winged predator, ready at any moment to rise from the chair and fly. He knew how to navigate on the tundra by the stars and how to use the intersecting wind angles carved in the snow as a compass. He knew how best to preserve fish in permafrost and when the bloom of certain flowers indicates a change in the taste of caribou meat. He knew how to follow birds and small animals to the animals he was hunting and also which birds would give him away to his prey, such as seagulls, which had become much more common than when he was younger. "They're always telling where I am to those caribou," he said. "That's why I don't like them."

Thomas easily summarized the differences he had seen in the environment over the last twenty years, simple highlights to represent the enormous changes he had observed. Open river ice in April—the first time ever this year. Riverbanks eroding much faster. Small ponds on high spots breaking through to the next lake or the river, something that began in the 1990s. Temperatures reaching 90 degrees F in the summer and warmer winters, with snow coming later and melting earlier. He used to be able to hunt with a dog team into mid-May. The rivers didn't break until mid-June. Now the rivers broke in mid-May or the first week in June. When snowmelt came, it was much quicker now. Mosquitoes arrived in the later part of June; they used to come only after July 4. The tundra bloomed earlier now, with more flowers. Hunting was no longer possible in June and July because the weather was too warm to keep the meat from spoiling. Far more seagulls and jaegers were flying in the area, and hawks appeared for the first time. Waterfowl were declining—maybe the hawks were killing them, maybe the foxes, which had overpopulated since fur trapping stopped—and some species had virtually disappeared. Foxes like to migrate out to the sea ice to follow polar bears, because the bears eat only the skin of the ugruk, not the meat, leaving plenty for foxes and ravens. That had happened much later as the sea ice was late in arriving. Some of the ravens overwintered inland now, guiding Thomas to find wolves.

He described the waterfowl and other migratory birds arriving in May. "It looks like lots of people are coming. They fill up the sky. It looks like lots of people coming in. They're singing and they fill up the air. That's the best time of year, when they come in May."

It seemed I had finally met a man living a real life. I greedily sucked in as much as he would tell me. But Flossie broke in, speaking Iñupiaq. Then Thomas became more cautious. "When white men find out they start doing what we do," he said. "So I have to keep it secret." Soon he got back into the flow, but Flossie spoke again. Thomas slowed down, noting that his elders predicted that one day he would be kicking whiskey bottles and dodging cars—and they were proved right. "That's what I learned from my grandparents. When I know where I am, maybe I better keep what I know about subsistence secret."

The start of the changes that white contact brought far predated Thomas Itta, Sr., but they continued all the time. For example, the televisions that were on nearly continuously in Alaska Native homes—they came with the oil wealth of the 1970s, when the state government installed improved communications in villages all over Alaska. As everywhere in the world, television was a wedge between young and old, a competitor to indigenous culture beamed as an ideal image from the world outside. When they had TV distracting their attention, young people were less interested in hearing stories or learning skills from elders such as Thomas and Flossie. Some absorbed the seductive allure of televised teenage rebellion, currently manifested by hip-hop culture, and adopted its attitude, alienation, and even baggy urban clothing—a choice that was laughably inappropriate in the Arctic. Thomas said some young people wanted to learn, but no one was teaching them. The elders were busy watching TV, too. When Thomas was a boy, he was prohibited from speaking Iñupiaq in school, even though he didn't know how to speak English. Today the school wanted the children to speak Iñupiaq, but they didn't know how. The constant bath of English via TV made it difficult for children to think in any other language. Less than 10 percent were growing up with Iñupiaq as the primary language at home.

These changes and the changes in climate interwove in Thomas's discourse.

"Our grandparents, our great-grandparents, they told us, Something will be happening in your future. And they're right. It's happening right now. They're gone now, but the change is happening now. Our grandparents told us, You will have everything in your future. Everything will start declining."

Even the snow changed. It came late this winter and the weather stayed warm, mostly above zero degrees F, until April. The snow on the tundra was thin and hard but in the bushy willows built into soft, deep drifts, as deep as six feet. The weather had changed, but so had the landscape and how it distributed snow. There never used to be so many willows. "They started growing here, there, and all over now," Thomas said.

When I mentioned what Thomas said about the snow and willows to Matthew Sturm, his eyes widened and focused on me with an intensity I hadn't seen before—for the first time, I had said something that he considered really important. Matthew and his colleagues began studying how snow and brush relate on the Arctic tundra more than five years earlier, working on the theory that they might form their own positive feedback loop. Dwarf birch, willow, and other brush trapped snow in drifts; the thicker snow insulated the ground; that insulation delayed the freeze and encouraged biological activity in the soil over the winter; and those effects encouraged further shrub growth. Matthew's original paper on the topic was rejected as too speculative, but he, Glen, and other collaborators kept working on the idea, measuring snow and bushes and modeling how increased shrubs would dramatically alter the distribution of snow. Snow would pile up in shrub-centered drifts, keeping the ground warmer and potentially accelerating decay and the release of carbon dioxide from formerly frozen organic material. The water content available to the ecosystem as a whole would be greater, because the shrubs would decrease the wind's sublimation of the snow by over 60 percent. If the phenomenon were real, it could affect the entire climate change equation.

In 1999, they had a big break. Matthew got word of a set of detailed aerial photographs of the North Slope taken in the 1940s. The U.S. Geological Survey had stored them for fifty years and was about to dump them. When he got a sample, Matthew knew he was on the verge of an important discovery. The photographs were extraordinarily detailed: the negatives were nine by eighteen inches. Scientists had made them to help the initial mapping of the region, a necessary precursor to oil exploration in the National Petroleum Reserve. Matthew and Ken found the men who did those expeditions, who took the photos and floated down the rivers, geologists mapping terrain while rafting its unknown waterways like modern versions of John Wesley Powell, the explorer of the Grand Canyon. Matthew chartered aircraft to photograph the same places again, re-creating as closely as possible the angles and distances. The side-by-side photographs told the story unmistakably. Brush growth had exploded across the Arctic. They quickly published a piece in *Nature*, with a pair of photographs, and were inundated with calls from the world media. Newspapers as far away as Africa carried the story.

Matthew's findings agreed with Thomas's on the 2002 transect as well. People in Barrow said it was a low snow year—the lowest ever, some said. In fact, the transect found plenty of snow had fallen, but wind had swept it all off the tundra, sublimated it, and piled it up in the willows, as Thomas observed. "I was at Barrow," Matthew said. "It did look kind of low. But the lines don't lie."

There probably weren't many other people on earth with the skills to perceive that difference. But Matthew and Thomas never met. When I told him about what Thomas had said, Matthew was in a rush, pushing to get out of Atqasuk on to another set of sites and on to the end of the transect, in Barrow. In his diplomatic fashion—aware of how his words could appear in print—Matthew had already expressed his skepticism about using traditional knowledge. The team spent time in the five villages along the transect teaching about their work in science classes in the schools, but they hadn't really tried to gather knowledge from local people. Younger Natives didn't spend as much time in the snow as Matthew did. They might go caribou hunting, but they traveled fast by snowmachine and then retreated to warm houses. How could they have much to tell him? Elders might know more, he reasoned, but probably only about issues that related directly to their hunting success. A city dweller might remember changes in his income but probably could not remember temperature patterns a year or two in the past. Matthew thought even the wisest Native hunters probably wouldn't know much about snow differences that didn't affect hunting.

Matthew was excited by the presence in Atqasuk of a fellow snow master, a real Native expert he could talk to as a peer. But stopping to talk to Thomas now was out of the question. It takes time to get an introduction, to build a sense of rapport, to make it clear you do not intend to steal the knowledge you seek. Even in the same language, you have to translate from one worldview to the other. You have to sit awhile. And Matthew was on the move. That next set of sites, equipment to load, the team ready to rush out the door and speed off on the machines. The irresistible drive of the expedition pulling forward toward its conclusion in Barrow like a rope drawn quickly across the face of Alaska and then gone.

CHAPTER FOUR

The Lab

AFTER A THOUSAND MILES, the cross-Arctic snow transect team emerged from the undifferentiated white plain around Barrow at a cluster of thirty-year-old construction camp trailers assembled into a T-shaped building. On the outside it looked like a big storage shed, except for the hand-painted words "NARL Hotel." I thought that was a joke the first time I saw it, then learned it really was a hotel, with small private rooms in units arranged on a wide central corridor, a coffee machine in the hall, and men's and women's shower and laundry rooms near the manager's office and living quarters. The NARL Hotel exclusively served scientists and visiting workers with business at the other NARL buildings, mainly the huge, H-shaped Building Number 360 that it hid behind, a one-story structure on stilts, with wings that contained BASC, Wildlife, and the borough's Ilisagvik College. At the meeting point of Building 360's wings, a modern cafeteria served meals to students, scientists, and, for Sunday brunch or prime-rib night, families driving out from town. The scientific parts of the building, flimsily built decades ago with thin plywood floors that rumbled underfoot and furnished in a 1950s office-drab style, had endured generations of abuse from Arctic weather and from workers encamped for short, intense stints of work. Along the walls in the BASC and Wildlife wings, clippings, scientific posters, photos, and project descriptions appeared but seemed never to go away; some, describing projects thirty years in the past, had almost faded out.

The team parked the snowmachines and checked into little rooms, each with its own TV. April Cheuvront felt the spell of the trip breaking, the transect team disintegrating one Alaska Airlines flight at a time. She had planned to visit Barrow, to buy Native crafts for her husband, and to see the Iñupiat Heritage Center, but she couldn't escape the strong gravitational attraction exerted by NARL's work camp culture and Matthew Sturm's single-minded work ethic. Most scientists and technicians visiting NARL saw little other than the cafeteria, the NARL Hotel, and the dilapidated shop or lab where they worked in the complex of seventy buildings. Some sat in the cafeteria under a big TV showing CNN or sports and told one another stories about the Eskimos in Barrow. Although it was only a ten-dollar cab ride or ten-minute snowmachine ride away, few went except to pass through the airport. Social gatherings in Barrow happened at home, at church, in camp—settings that didn't include strangers. A group of young Australian technicians working on a robotically controlled aircraft called the Aerosonde got together on their rare day off to "go into town" but came back before long. There were no public buildings other than the pizzerias and diners, the Heritage Center, the big store Stuaqpak, the airport, NAPA, and the gas station. There were no bars, as the sale of alcohol was illegal. It was more fun to stay in the big Quonset hut workshop—known as the Theater, since the faint outline of that word was the only marking left on the snow-scoured exterior—and work on the airplanes, as they did every day of their six-week tour.

The transect team members who remained mostly kept to themselves in the evenings, since the rooms were too small for more than three people to sit, but they still ate together in the cafeteria and worked together on the snow. Matthew had more sites to do in Barrow, more multiyear data sets to maintain with hundreds of measurements. Glen Liston and Jon Holmgren were gone, but April, Ken Tape, and Eric Pyne remained to work the long hours Matthew pushed them to, although never as long as he'd have liked. Ken hung his head in the cafeteria and said, with a smile, "When you guys are gone, I'm dead."

Matthew was working on several projects at once. Like many Arctic field researchers, he funded his work entirely with grants from the National Science Foundation or a few other agencies. Matthew was a moneymaker for the Corps of Engineers, because each grant covered not only his salary and expenses but also a generous overhead percentage—more than enough to keep his Fairbanks office heated. Grants typically lasted three years and covered only a portion of a researcher's time, so soft-money researchers needed several projects at once to keep themselves and their labs busy and to provide consistent work for their teams of technicians, students, and postdocs. Like small businessmen, they needed constantly to be on the lookout for funding opportunities that matched up with the projects they already were doing to make the most of time, staff, and travel. Matthew had mastered the juggling, but it wasn't easy for anyone else to keep track of his overlapping projects and side interests. The team followed him each day and took the measurements without worrying too much about which project they were working on.

Spring whaling season hadn't started yet—Oliver Leavitt was still building his boat in the Heritage Center—when Jim Maslanik, a grad student, and a postdoc from the University of Colorado at Boulder arrived to join Matthew and his team for yet another project. Jim checked into the NARL Hotel, as he had done many times before. A soft-money scientist, like Matthew, he spent a lot of time away from home fulfilling cold-weather research grants and attending scientific meetings to advise the seemingly infinite number of committees that set priorities and goals for the grant-writing research community. He desperately missed his teenaged children back in Boulder. "If your kid's fourteen, that leaves a few more years and they're off," he said. "I regret every minute I'm not there."

Jim's project was to bridge different scales of seeing sea ice, as Matthew had been trying to do for snow on land. His specialty was remote sensing of Arctic ice—mainly via sensors on satellites—but he also had developed projects with other ways of looking at ice, which he pursued with intensity as he grasped the vast gaps in knowledge about the Arctic system. While Matthew dined with gusto on the unknowns in Arctic science, Jim saw hungry voids. It seemed to him possible that climate science could not answer the questions society wanted to know. Many scientists were pushing, studies were becoming more detailed, models were becoming more complex. But the barrier to real understanding stretched like a rubber membrane—maybe they would never break through.

"If we could step back a few strides, if we had the brain power to just observe the system, like a neural network, and not necessarily understand but see what happens and get some kind of larger understanding," he said. "In the big scheme of things, what does it mean? Here we are scrambling away for every detail, and someone in the future may look back and say, 'These guys sure worked hard, but they were on the wrong track.' And if you get that kind of thought, it gets hard to do science. I don't think I could get excited just to learn to understand snow the way Matthew can."

Matthew didn't take such pessimism seriously. He said Jim always talked that way, but he kept on working.

For this project, Jim had responded to a call from NASA for proposals to validate images produced by a new satellite called Aqua, an element of the international Earth Observing System, which theoretically would be able to measure the temperature, thickness, snow cover, and microstructure of sea ice, as well as many other things—evaporation, water vapor in the atmosphere, phytoplankton in the ocean, soil moisture, and so on. A satellite sensor called the Advanced Microwave Scanning Radiometer for the Earth Observing System (AMSR-E) would look downward with eyes that could see in the microwave band of the electromagnetic spectrum. All matter warmer than absolute zero emits radiation in many frequencies, called black body radiation, the most familiar of which is radiated heat, or infrared radiation. When you look at something with light, radar, or sonar, you have to actively bounce a signal off it and pick up the reflection; this microwave technique instead passively picked up the signals everything constantly put out. For the purposes of studying sea ice, the AMSR-E could measure the extraordinarily weak microwave signals emitted by water in each of its forms. The strength of the signals varied in different frequencies depending on whether the water was ice, snow, or liquid, salt or fresh, its temperature, and its underlying composition. This would not be the first passive microwave sensor on a satellite—others had been returning data since the early 1970s—but it would be an attempt to deal with large inaccuracies in earlier efforts, especially in recognizing ice of different kinds and thickness, and in distinguishing between open water and melt ponds, water sitting on top of the ice, which constitute up to 20 percent of the ice surface in the summer. Passing by 710 kilometers above the earth, the satellite would return loads of new information. But the job of figuring out what the measurements meant would remain.

When you look out on a sunny day, your eye does only a small part of the work. Your brain automatically translates patterns of energy in the frequencies of visible light into a picture. Your experience of how different materials look helps you recognize what the picture represents. The analogy isn't exact, but a sensor such as the AMSR-E starts out like an eye without a brain. It receives microwaves but doesn't know how to make them into a picture or how to recognize what it sees. Building the software to give it those capabilities was hard. The AMSR-E produced strings of numbers for two polarizations of microwaves in six different frequencies, and from that the computer needed to recognize at least a dozen kinds of ice, including ice with water on top of it, and the other properties embedded in the information. A single pixel was between five and fifty-six kilometers on a side, so one point could contain many kinds of ice and water. By the time the satellite launched in May 2002, Japanese and U.S. teams had been working for eight years to develop computer algorithms that could translate the flood of numbers from the sensors into grids showing ice types and conditions on the surface (as well as rain, clouds, sea surface temperatures, wind, and other variables). But they knew the algorithms were imperfect. In 1998, Jim and his colleagues had used three leading algorithms to measure the same patch of ice and water around a frozen-in icebreaker: the concentration of ice in the water came out anywhere between 62 and 96 percent.

Jim's problem was how to square what you could see and measure on the ice by hand with what the satellite could see. Two dozen other teams would be doing the same kind of work all over the world, in Australia and Africa and the Chesapeake Bay, among other places. But most of them were working on big, homogenous areas that would stay still. On floating Arctic ice, conditions were different every few steps, and the whole picture could change in a day. Jim's idea was to capture the scene by stepping down in scale, zooming in from the wider to the narrower, at one place and time. The satellite would look down from space; a four-engine NASA P-3 aircraft would look down from as high as twenty thousand feet; a robotic Aerosonde aircraft would look down from lower and with finer detail; teams of scientists would look down while standing on the ice, making measurements and drilling cores to examine later under a microscope; and an Iñupiaq elder would come along to say what he could see. All these views would be coordinated so that everyone was looking at the same ice at the same moment. If it worked, it would be a logistical masterpiece, an abstract idea about different ways of seeing, played out on the stage of reality.

Matthew and Jim met in Jim's room at the NARL Hotel, crouching on the bed, to plan their dry run—the real show would take place a year later. With his great experience in the field, Matthew pointed out a lot of ways the plan could go wrong; the purpose of this field season was to find the pitfalls. For example, when Matthew had tried previously to match up ground-based observations with aircraft, the planes never seemed to show up in the right place—they said they were on the coordinates, but from the ground he could see them way off to one side, with no way to get them to where they belonged. Jim had done this kind of work before, too, and he was having special beacons built that would pin the planes to the people on the ice. Together they inspected a recent image of the ice around Barrow made by a satellite using Synthetic Aperture Radar, called SAR, and tried to pick out sites they would like to visit, white lumps and gray patches on the SAR image they could go to on the surface to find out what they were. When the real project started, Jim hoped to use fast copies of AMSR-E images to help his team scramble to the sites of interesting features, an idea Matthew approved of. He said, "Usually what's happened in the past is we get back and get a satellite product and somebody says, 'What's this white spot?' and you say, 'I don't know, we didn't look at that.' " They also talked about using the advice of the Iñupiaq elder, who could be questioned on the spot. "Look at the SAR image—he may know what a lot of this stuff is," Matthew said. Jim planned to keep careful records of the elder's comments on ice conditions and what created them. "We have to get our knowledge base up," Matthew said. "That's more important than getting a set of transects."

Matthew was bringing up all the potential problems, but he was optimistic. "This is going to be fun. I'm psyched." Jim, who had already prepared for many pitfalls, hung his balding head and shoulders like a losing baseball coach. He said, "The important thing is that we fit everything in. If we miss one of those scales, then we miss the whole thing."

Craig George had recommended Jim and Matthew hire elder Warren Matumeak. Warren had been a whaler and polar bear hunter, as well as a borough planning director and cofounder of a traditional dance group, but at seventy-four he wouldn't be tied up with the whaling season. He had been raised in an uninsulated shiplap house by a father who lived the traditional, subsistence way, and he had begun his science career as a child, trading rare plants for hard candies with the original Charles Brower, who sent the specimens to museums and botanical gardens around the world. Jim had found Warren uncommunicative the first few times they met, and was afraid he had insulted Warren by trying to lowball his consultant fee before negotiating pay of a few hundred dollars a day. But when I met with Warren, money didn't seem to be the issue. He was simply skeptical of the scientists' ability to learn much with satellites and computers. He said satellites were useful to show where ice was moving offshore, which helped warn hunters to come in before a collision. But he thought the complexity of ice, its layers, movements, and changes through time, were too subtle and complex for science to predict, and probably beyond the limits of human understanding. Meanwhile, the scientists' knowledge was rudimentary. "I think they've got a long way to go," he said. Like other elders, he had already spent years telling scientists how things were so they could find out for themselves. "I've been telling them, this is how it is. They go do scientific study and do a lot of work to prove it, and they come back and say, 'Warren, you were right.' It's just common sense. They use science to prove things we already know."

On the other side, Matthew and Jim wondered if they could develop valuable communication with Warren. Matthew had already told me his reservations about using traditional knowledge in Atqasuk; the generation gap between the scientists and Warren would only add to the cultural gap. Jim had started as a believer in using Natives' knowledge and even served on BASC's science advisory board, but now he worried about translating the different frames of reference. When the scientists wanted to know how old a piece of ice was, Warren talked about how fresh it had become, not how many years it had been around. It was the problem of complexity. The physical scientists wanted to know irreducible facts, but Native knowledge was tied up with experience. You could try to strip away the experience to get at the facts—parsing out a hunting trip to get times, places, and events, for example—but the complexity seemed never to recede. Jim said, "Every time you get a layer of it, there's another layer. And with each layer you go down, it gets less definite."

Jim, Matthew, and their team met with Warren at the BASC warehouse, where the Colorado postdoc and grad student working with Jim were given snowmachines and cold weather gear. They put on BASC's ratty red parkas, which had patches that said "United States of America Antarctica Program" and the notation "2nd" in black marker; these were hand-me-downs from the better-funded South Pole program. The white people added goggles and face masks to cover every inch of skin, but Warren mumbled, "You guys cover up too much," and took off on his snowmachine directly into 10 degree F weather and a 30-mph wind, his face uncovered except for the metal-rimmed glasses sitting on his nose. The team flew over the flat ice of Elson Lagoon, which is protected from the ridging pressure of the pack ice by the hook of Point Barrow and the barrier of the Plover Islands. They stopped to inspect a crack where the snow had crusted over into a shiny surface. Matthew asked Warren what caused it and he said, "You know as much as I do, I'm learning, too." In the excitement of his curiosity, Matthew bent down and then crawled, tasting the snow and quickly dreaming up a theory about how wind had pumped seawater through the crack and spread a coating of salt over the snow. After a calculated pause, Warren said, "You got a good imagination. I never seen anything like that." Matthew's face turned as sour as if he had hit a rotten bite in an otherwise delicious apple. Without further comment, we rode on.

The wind was kicking up the snow and making navigation by distant visual landmarks impossible. The team carried GPS units programmed with waypoints based on the SAR image; the GPS would lead the snowmachines to these preselected sites of interest, a first test of the process of coordinating different sets of viewers on the same features. But it was not easy to convert a satellite image into simple latitude and longitude coordinates, and something had gone wrong—the GPS points all seemed to be skewed. After puzzling over the GPS displays in the sharp wind and blowing snow, Jim and Matthew gave up on them. They would have to do the math again and reset the waypoints indoors. Instead, Warren led the way. First he drove dead on to Plover Point, the far eastern tip of the hook of Point Barrow, then led the group to a big island of multiyear ice that showed on the SAR image as a white blob, more than four miles east of the point. Matthew said he knew how Warren did it, using the wind direction as a compass and setting a course by his memory and sense of direction, but I was impressed—there was nothing but chaotic whiteness out there that I could see.

The ice stood in a crazy jumble near the huge multiyear berg, which,

Warren said, had grounded and dammed up the movement of the first-year ice, piling it like a rucked-up rug. Matthew said, "How do you measure this?" There was so much variability—where would you start? The multiyear ice loomed above like a big white bluff, twenty feet high, rounded into voluptuous shapes and glowing neon blue where white snow and ice shavings had been scraped away. Team members climbed around like discoverers of an ancient ruin. The postdoc climbed to the top of the highest round hill and gazed out in the sunshine on endlessly varying discontinuities in the broken ice. It would have been easier to take in as pure chaos, but at every scale there were textures and patterns being broken and superseded. He was from Australia and had never seen sea ice before, but as a climate modeler he had tried to reproduce its effects on a computer. "This puts my models to shame," he said. Everything we could see would fit in a single grid square in a global model, where it would all be represented by a single average. "Coming out here is really exciting, but it's also really depressing," he said.

Any complete model, or even a solid theory, of how the Arctic affects global climate would have to take into consideration ice albedo and heat transfer, known as heat flux, from the Arctic Ocean to the atmosphere. Farther south, relatively simple equations might be able to capture the sun's heating of the sea (leaving aside currents, salinity, temperature layers—it was never really simple). But with ice in the calculations the problem became horribly complex. Snow on ice reflected 80 percent of the sun's energy, bare ice 65 percent, melt ponds on top of ice 35 percent; open water reflected only 7 percent, absorbing 93 percent and warming accordingly. Besides reflecting the sunlight, the ice insulated the boundary between the ocean and atmosphere. Where the ice was solid, little heat transferred. Where open leads formed, enormous fluxes of energy could pump upward from a relatively warm ocean into air that could be far below zero, heating the air, producing fog, and even powering storms. The sizes of ice pans and leads influenced these effects—it was not a straight-line relationship of more water equals more heat flux. And it changed daily. As ice melted through the spring, the size and shape of leads and melt ponds and the reflective quality of the ice and snow surface changed in a complex, swirling dance.

Reduced sea ice seemed likely to be a powerful accelerant of a warming climate, as the positive feedback of decreased reflectivity and widening leads contributed to warmer atmospheric temperatures, further decreasing albedo and melting even more ice. But the process was far from simple and large gaps remained in understanding it. Although Arctic sea ice was quickly thinning and decreasing in extent, atmospheric circulation and ocean currents played a hand: some of the ice simply floated north or away into the Atlantic. Some calculations suggested the decrease should be even larger. And despite recent ice years that were the lowest in history, in 1996 there was a very high ice year. One theory held that climate change had switched on an unknown system of extreme ice variability along with reductions in ice.

It would take some basic information to resolve these questions. We still didn't know how much solar heat ice takes up in the summer and could not predict summer albedo. There was a lack of data on first-year ice in the summer—it was too dangerous to work on—but that was where much of the melting that drove the feedback loop took place. Surface data came from a few ice camps, buoys, and icebreaker cruises. Satellites were the only real hope for a broader picture. But satellites using visual light worked only in the daylight and when the sky was clear, a small fraction of the year. Besides, the ice pack looked like a confusing mess from space, even using radar—it was difficult to tell different types of ice apart or to distinguish melt ponds from open water. Combining passive microwave data with other satellite information could be the key.

The team stopped for lunch behind an ice wall that concentrated the sun's warmth and provided shelter from the wind, then set to work, while Warren silently scanned the horizon for polar bears. Ken set off on a snowmachine with a GPS unit to circle the multiyear island, drawing a digital line around it to get its size. Matthew sighted through the theodolite while Eric, scrambling over the jagged terrain, set up a mirror for the instrument to register on at points along an imaginary line into the distance, precisely surveying the ephemeral landscape for a statistical measure of its roughness. Another group drilled ice cores, taking temperature readings in the holes at intervals going downward. They registered lower temperatures going deeper. Warren questioned that with a skeptical look and a shake of the head—since the water below was warmer than the air above, the ice temperature should rise as the drill bored downward. Matthew speculated the ice remained chilled from a past cold snap. Then he found a problem with the instrument; in fact, Warren was right. Jim tried to reach the Aerosonde crew on a radio to bring in an aerial perspective on the site, like a platoon leader calling in an air strike. But he couldn't reach them.

On another day, I stayed back at NARL to see the Aerosonde guys' perspective on the project. The planes looked like large hobbyist models, with wingspans of about ten feet and weight of about twenty-five pounds. Their advantage lay in their ability to take measurements for long periods of time, low to the ground, in places too dangerous for manned planes—the name Aerosonde was a play on radiosonde, the technical name for weather balloons. In 1998, one of these little planes became the first unmanned aircraft to cross the Atlantic Ocean, taking about twenty-seven hours and using one and a half gallons of gas. With NSF funding, Jim and his colleagues at the University of Colorado were working with Australian Aerosonde Ltd. to adapt and use the plane for remote sensing in the Arctic. The Australian crew also had spent long stints in Florida, flying the planes into hurricanes, and in Japan, Korea, and Micronesia—they were climate change research soldiers of fortune.

To make an Aerosonde airport, BASC workers had dragged a little building on metal skids to the abandoned NARL runway, a mile-long strip of metal plates north of the main compound. Team leader Maurice Gonella called off checklist items to Dan Fowler, verifying communications, controls, the infrared sensor—it sounded like getting ready for a moon launch—and with all systems go, Dan set the airplane in a cradle on top of a pickup truck and started its tiny engine. The pilot, Dennis Hipperson, wore a radio control with little levers like a weekend hobbyist. As Dan drove the truck down the runway, into the wind, and Maurice monitored the engine on a laptop in the shed, Dennis pulled back on the stick, flipping the plane aloft like a paper plate in a strong gust. Soon it was a speck up above and Maurice took over control with the computer—Dennis's job was done until the plane returned and needed to land. Maurice sent messages to the airplane's onboard computer specifying locations and altitudes, and the plane went to those spots, sending back a steady stream of data, which at the moment included temperature, altitude, humidity, air pressure, and, through the infrared sensor, temperature of the ice down below. The Aerosonde was supporting a Department of Energy atmospheric observatory as well as Jim Maslanik's project that day.

Jim, out with his field team, called in by cell phone. Maurice's side of the conversation sounded like this: "East side of the spit? What spit is that?

Where the road stops? Isn't that west of the point?" Using the map on his computer screen, he outlined a box for the plane to fly around somewhere over Jim's team, but it seemed rather vague. From the display, there was no way to tell if the jumping infrared temperature readings matched the particular spot where Jim was working. When I asked Jim later, he confirmed he had never seen the Aerosonde overhead, but then the plane was so small it was difficult to see even when you knew where it was.

With each setback, Jim and Matthew congratulated each other. NASA had tried to cut this dry run from the project budget, but the scientists had said they wouldn't do the work without it. Attempting to coordinate crews and expensive aircraft and satellite time without first learning the sheer difficulty of it could have amounted to a fiasco. Instead, the measurements they were taking were simply practice and the problems that arose provided opportunities to devise ways of getting it right when it mattered.

Jim's intention to take copious notes of Warren's comments was one of the elements of the plan that proved unnecessary. At the big multiyear berg that first day, Warren didn't say much, didn't join the team for lunch, stood off staring at the horizon—still scanning for polar bears, I knew, but it looked more like lonely meditation. He led the scientists where they needed to go and kept them safe, but more as a guide than as a participant. Nor did I see the scientists engaging him often with questions or observations. His understanding of the ice in all its dimensions, from the atmosphere to the sea floor, ocean currents, and development through time, was the view the scientists were ultimately after, but they would approach it only years and millions of dollars from now. At the moment, Jim and Matthew were looking for numbers a computer could understand—the height of an ice chunk, the temperature of a layer, the size of a grain—and what Warren had to say about the complex mechanics was almost irrelevant. As Warren told me, he didn't know how to operate a computer, just knew that although they are supposed to work very fast they seem to take up a lot of people's time. And I didn't think a computer could understand Warren, either.

We left the multiyear berg heading westward, guided by Warren, to see more of the ice shown on the SAR image and pick out sites for study on future days. West of Point Barrow we came across ABC Crew's whaling camp, set up in the lee of a pressure ridge against the windstorm. By toughing out the wind here, the Browers would be well ahead of all the whalers waiting back in town, with a good chance of getting the first whale. But first they would need a lead of water to put their boat into. West of the camp, the ice stood up in row after row of high, steep ridges, relatively newly crunched together into a virtually impassable series of walls.

Warren pointed out the water sky in that direction. Water sky: where a lead of open water darkens the sky above it. To the west and south, beyond the ridges, a strip of the sky above was darker, an amorphous zone where much of the light had been sucked out of the blue. That was where the pack ice separated from the shorefast ice. The liquid sea was swallowing the sunlight, leaving that much less to illuminate the sky. Here was the physical manifestation of what we had been talking about, the ice albedo in the flesh. Over the ice, the sky was bright blue with reflected light; over the water, the sky turned darker blue, a strange, stormy blue in contrast to the brilliant color we were used to but, I suppose, the normal sky blue of the ice-free temperate and tropical part of the world. That darker blue represented a deficit in light created by the water's five times greater absorption of energy compared with the ice where we were standing. The weirdness of it made me think of a rip in the sky's dome through which we were seeing another latitude or another time, a place or time without ice. Normality lay in the white surface and bright blue sky of the ice world; that dark rip seemed vaguely threatening.

Eric climbed to the highest point on the highest pressure ridge and I followed him. From a precarious perch up there, we could see a thin black strip of ocean far in the distance, its waves appearing as fleeting silhouettes against the white of the opposite side of the lead. I could have been imagining, but I thought I saw whales.

One fine summer morning decades ago, John Kelley, a marine scientist and later director at NARL, came to Kenny Toovak, who managed the lab's boats and equipment, and asked for a ride out to Point Barrow in one of the eighteen-footers with an outboard motor. He had work to get done and limited time and wanted to go right away. Kenny looked at the sky and told him, with typical Iñupiaq indirectness, "I'd like for you to wait a bit."

John didn't insist at first but paced around impatiently, making it clear he needed to go soon and saw no reason not to. The weather looked perfect. In fifteen minutes, he came back and told Kenny it was time to go.

Kenny, a skilled storyteller who knew how to draw out every detail in a slow, dignified style, said he told John Kelley, "You really want to go out, I'm going to give you a boat and an outboard. You can go. But I'm not going to give you a driver. And I don't think we're going to look for you even. You really want to go out, go on and go."

John returned to his office. Shortly, the wind picked up. Soon it was howling, with white caps frothing on top of the waves. He returned once again and said, "Kenny, I thank you for not sending me out."

Anne Jensen didn't believe the Iñupiat had ever been proven wrong about a major environmental issue. She said their precise language and well-trained memories made them reliable sources over centuries. The elders knew that a ruin at Utqiagvik was a qargi before she and Glenn dug it up and confirmed it had that purpose. It hadn't been used in five hundred years. The elders' ability to predict the weather could be valuable if scientists could reverse engineer it and find out how it worked. "They must be using physical correlates," she said. "We have to figure out what they are."

But it's not easy to dissect the magic of what an old man feels in his bones. When I asked Kenny what he saw that day decades ago he said, "It was something about the sky, the clouds and south wind, a bit warm. It's always kind of rapid, it always happens in a rapid way. I learned that lesson from my parents and from the elder people. When the wind is kind of blowing from the south you better hold off for a while and see what the weather will do."

Elders across the Arctic said that the weather had become erratic and more difficult to predict since the climate started to change. Atmospheric scientists agreed that the weather was more changeable and cyclonic storms more frequent in the Arctic with the other changes of the last decade or two, shortening times of stability and perhaps breaking the rhythm of the winds that the elders had learned to anticipate.

I asked others for similar stories and whether they agreed. Some gave examples, but more—including Oliver Leavitt—gave me versions of a favorite local story that goes roughly like this: A crew of white scientists were stationed at a remote camp to do research with an Eskimo as their guide and laborer. The first morning of their project the guide stepped outside during the breakfast hour and took a short walk, coming back in to say the weather was threatening and they better stay in that day. Sure enough, a violent storm kicked up. The next day, he took another stroll during breakfast and reported back that conditions would improve. The team had a successful day in the field. This went on for some time and the scientists came to rely completely on their guide's uncanny ability to predict weather conditions. One day as he was leaving the breakfast table one of the scientists asked to go along so he could learn to divine the weather, too. The Eskimo said, "Sure, I'm just going over to my tent to listen to the seven o'clock forecast on KBRW."

The joke was on me. My question was irrelevant: modern weather forecasts were far more reliable than the old way. People still knew the pattern of the winds, which tend to wheel around Barrow counterclockwise with the passage of circular weather systems. Whalers also knew how the pattern of winds affected safety on the ice. But the National Weather Service was better at predicting the future than they were. And its technology appeared just as magical to anyone who didn't understand how it worked.

Kenny Toovak's life had been dedicated, in various ways, to bridging the gap between the races that created the illusion of magical powers on either side. He was born when official government policy was to suppress Native cultures. He became an adult when some businesses in Alaska cities still posted signs saying No Natives or Dogs. His father was a successful hunter and whaling captain, a man of few words who had taught him never to put himself higher than another person. Kenny never had his own whaling crew, for which he felt shame. Instead he supported his family in the beginnings of Barrow's government-funded cash economy. As one of the first Natives to hold a paying science job in 1946, he drove a dog team and handled a lead line in a small boat to help the U.S. Coast and Geodetic Survey map the area around Barrow. Money allowed him to buy fuel oil to keep warm instead of burning seal oil and blubber, a change elders told me marked the greatest improvement of their lives, the ability to get away from the cold, and helps explain why their houses are kept so hot today. Although hunting with his father allowed Kenny to attend school for only six years, he eventually oversaw crews of science support workers at NARL. He had followed a courageous path. Some Alaska Natives responded to racism with hostility toward whites and contempt for fellow Natives who acted too white, but Kenny never kept aloof. He believed in science, believed in unity and understanding between the races, and built friendships with scientists whose careers he boosted with little acknowledgment in return until, in his old age, he was honored by scientific gatherings as a major contributor to Arctic research and an expert on sea ice.

Kenny started his working life knowing nothing about whites, their customs or food or what they expected from him on the job. One morning early in his career he woke in Dease Inlet, east of Barrow, on one of two sixty-five-foot landing craft with a crew of white men and two other Iñupiaq workers, Tom Brower, Sr., and Roy Ahmaogak, Jr., where they had stopped for the night before heading upriver toward a drill site (this was the same navy oil exploration program that produced the aerial photos that Matthew Sturm was so glad to find fifty years later). Kenny was the youngest on board, so he volunteered to cook breakfast, taking the order of a Schlumberger mechanical expert who was a passenger: two hotcakes with fried eggs on top. He figured out how to make batter from the pancake mix, lit the Coleman stove, and poured two hotcakes in the pan, then broke two eggs, one on top of each uncooked hotcake. "I understand that's what he wants. So I look at my cooking there, and boy, there was something I should do better." He slid the mess overboard and tried again until it looked right. "I was scared of white people," he said. "I stayed to myself. Pretty soon, those folks started coming around, asking questions. I started getting to know these people I had been scared of. We started to share the different things we know. This sharing became a very important part of my life." But the food on that particular trip didn't get much better. The same evening, Tom Brower was cooking dinner when he called Kenny over and said, "I start cooking this steak, but boy that fat that I used wasn't really satisfactory. Smell." It turned out he had used dish soap.

Over the twenty-three years of the navy's operation of NARL, three hundred Iñupiat worked there, many selected for their skills by Kenny. Max Brewer, director of the lab from the late 1950s to early 1970s, who was called "King of the Arctic," relied on Kenny as a close adviser and said his clout was third only to his own and that of his assistant, John Schindler, who later became director. Kenny and the Iñupiaq workers he recruited to the lab taught nature and survival to generations of scientists. Max wrote, "In those days, if a scientist wanted to research long-term animal cycles on the North Slope, or the earlier snow geese populations in northern Alaska or on Banks Island, he consulted with Tom Brower, Sr.; if interested in fish populations and ranges, he consulted with Arnold Brower, Sr.; if he sought to develop realistic dioramas for displaying bird and animal specimens, he consulted Harry Brower, Sr." For the first half of the twentieth century, most of the Arctic biological specimens held by American museums were collected and prepared by the sons of Charles Brower. George Leavitt, Oliver's cousin, worked for the National Bureau of Standards for seven years in the 1960s operating a camera that measured the ionosphere. When the station closed, the bureau asked him to move to Boulder, but that was inconceivable to him and he stayed in Barrow, retiring from the BASC warehouse in 2002. Mostly, scientists used local Natives as guides to operate safely in an unfamiliar environment. Kenny helped with the science when asked, but other times he didn't even know the purpose of the studies he worked on.

The relationship was paternalistic, according to one scientist—why didn't Kenny become codirector?—but in the context of the times, Kenny had extraordinary power to aid his project of cultural bridging. He used another food story to illustrate. Kenny was leading a large party overland from Barrow to the Colville River, a distance of about 150 miles, with Native support workers, three scientists, and various members of his own family. At lunchtime, one of his female relations said to him, "We'll have to make two tables, feed the white people's food and then our own food." Kenny said, "No, we'll all be one family." They served only Native foods. The scientists were tentative at first but within a day or two were eating their maktak, heart soup, and caribou with gusto.

From such starts, the Iñupiat, who had never been conquered by war or treaty, eventually took control of their region—the land, political institutions, economy, education, and churches. Oliver Leavitt, a member of the generation that followed Kenny Toovak's, became one of the most powerful men in Alaska. In the early 1970s, he helped guide the creation of the North Slope Borough and the Arctic Slope Regional Corporation. ASRC was one of thirteen regional corporations and more than two hundred village corporations that received the settlement payment of Alaska Natives' land claims in 1971, when Congress ceded forty-four million acres of Alaska and almost $1 billion (the corporate system was chosen as an alternative to creating tribal reservations). ASRC went into oil field services on the North Slope and acquired subsidiaries all around the United States, becoming the largest corporation in Alaska, with five thousand employees, five million acres of land, and limitless political clout. Oliver was chairman. Politicians feared and courted him. When he held political fundraisers in Anchorage, ASRC suppliers and would-be suppliers came out of the woodwork with checks. Ted Stevens, the chairman of the Senate Appropriations Committee and third in line of succession to the presidency, called Oliver a close friend; Ted's late wife, Ann, had been an annual guest in Oliver's whaling camp. There, Oliver told political stories like hunting stories. In his tales he wielded Washington lawyers like rifles, choosing those with access to key committee chairmen, and his quarry was the technical amendment, a few words or a bit of punctuation that could yield a rich harvest. One such amendment in the 1986 Tax Reform Act brought Alaska Native Corporations $400 million in cash at the expense of the federal treasury.

Yet Oliver remained Iñupiaq; he used power mostly for his people, not himself. In 2002, after he cochaired the successful gubernatorial campaign of U.S. Senator Frank Murkowski, Oliver consented to be on a published list of candidates for appointment to the vacated Senate seat, but he wasn't really interested. He preferred his little white frame house with a tundra yard, eroding on the sea bluff, the quanitchaq entryway dark and cluttered like others in Barrow—a house roughly equivalent to a lower-middle-class starter home in most American cities. Showing off wasn't good manners in Iñupiaq culture. (Early in his career, Oliver's possession of three-piece suits, Super Bowl tickets, and a Mercedes had made the newspaper in Anchorage, but I saw no extravagance in Barrow.) As a whaler, he opposed offshore oil development that could threaten the bowhead. He strongly supported oil drilling on shore—that was where all the money came from—but in the spring of 2002, when Ted Stevens called to ask him to go to Washington to lobby the Senate for the opening of the Arctic National Wildlife Refuge (ANWR) for oil exploration, Oliver stayed in Barrow, hurrying to finish his new boat at the Heritage Center. "We should have done this over the winter," he said. There had been too many distractions. "It's a lot of goddamned work running a billion-dollar corporation."

Instead, Oliver sent to Washington Richard Glenn, ASRC's thirty-nine-year-old vice president of lands, a geologist and a whaler with natural talent in politics that Oliver said was second to none. Richard knew politics, but Kenny Toovak's dream of connecting science and Iñupiaq knowledge seemed to interest him more deeply than Oliver's more finished project of extending Iñupiaq political power. When NARL closed in 1980, the village corporation, UIC, saved the buildings, which the Defense Department deemed best suited for the bulldozer. Research continued at a low level, with UIC providing much of the support, until Richard formed BASC with funding from the National Science Foundation. He put Glenn Sheehan in charge as director because of Glenn's commitment to research that truly involved Native people.

Various scientific and cultural institutions had developed ethical standards for researchers gathering and using traditional knowledge, but the standards were often ignored or used only as window dressing, according to Patricia Cochran, director of the Alaska Native Sciences Commission at the University of Alaska Anchorage. It wasn't a problem only in cultural studies: researchers in the natural sciences exploited local people, too. Control of science logistics and the old NARL facilities would give Glenn Sheehan the leverage to demand ethical practices from researchers and to involve the Iñupiat in planning, hire them when possible, give them credit for their contributions, and report back to the community the outcome of studies.

At least that was the idea. Before the whaling season in spring 2002, word came that a scientific icebreaker cruise would be coming through at the same time as the whales. Senior whalers were furious, convinced the noise of an icebreaker would disrupt the migration and the hunt. Unsatisfied with the response from the organizers of the cruise, George and Maggie Ahmaogak fired off a letter to Ted Stevens challenging the entire program. The letter soon found its way down to the NSF's Office of Polar Programs—the people funding both the cruise and BASC. Glenn was on both sides and in the middle.

Jacqueline Grebmeier, a brilliant University of Tennessee oceanographer, had organized the cruise. Through her academic career, moving from project to project, Jackie had managed to get on Arctic seas fifteen straight years by putting together one grant after another, catching icebreaker "cruises of opportunity," and gathering her own long-term, standardized measurements, the kind that impatient institutions rarely fund. With these measurements, Jackie and her husband, Lee Cooper, also at Tennessee, developed a picture of how ice algae and other ice-bound organic matter falling to the seafloor during the melt helps power the biological productivity of the Arctic continental shelf, nourishing eider ducks, walrus, whales, and other bottom-feeding higher organisms. With less ice in recent years, the biomass on the sea floor appeared to be declining. Jackie saw herself how bivalves that once measured three inches now were only an inch in size. Meanwhile, the population of spectacled eiders in the area she studied dropped by 90 percent.

Now, finally, Jackie had the funding and institutional support to take this work all the way to developing an understanding of the entire system at the edge of the Arctic continental shelf. She would lead a ten-year program with a broad-based team to look at all aspects of the ice-melt zone, including oceanographers, biologists, ecologists, seafloor and sea ice experts, and so on. The first icebreaker cruise would leave from Nome with thirty-five researchers onboard representing forty Principal Investigators. She seemed to have thought of everything, even a special preparatory meeting to build rapport among the science team's members so they would be ready to collaborate as soon as they got onboard. Jackie was a remarkable woman: cordial and generous with her time, but also with a hard, sharp blade of intelligence, speaking fast in dense, technical language, her thoughts moving forward too aggressively for her to worry much about who could follow. As the cruise approached, she glowed with the accomplishment it represented: she had been working on it since before her daughter was born, and her daughter was seven years old. Then, suddenly, came the word that Barrow's whalers were up in arms and using their political connections to shoot down the entire program.

The scientists' lack of sensitivity made Glenn mad and he wanted the cruise canceled to demonstrate respect and rebuild trust. From the perspective of Washington, however, the Barrow whalers appeared hypersensitive and spoiled by their easy access to Ted Stevens. Richard recognized a compromise was in order—it was a big ocean—and he started talking. One morning he casually mentioned over breakfast that the icebreaker issue had been settled and sketched the ship's new route on a napkin. A week or two later, everyone knew.

In promoting BASC, Glenn tried to sell the Iñupiat as an asset. The old NARL facility had little else to recommend it. Originally a compound of 191 buildings with a 5,000-foot runway and as many as 220 year-round employees—by far the largest lab of its kind in the world—NARL today looked like a collection of ramshackle sheds and abandoned junk. During the spring and summer field season, scientists from all over the country filled labs inferior to middle school science classrooms. There was a shortage of snowmachines and other field equipment during the best periods for winter work. Just checking e-mail on the antiquated computers and unreliable dial-up connections in the work areas was a laborious chore. Antarctica and other polar research sites had more funding and better facilities. But Barrow had people. "Almost any adult you meet on the street who is over 35 [has] worked personally with scientific projects and scientific teams and can critique a field plan, and usually improve it," Glenn said. "They do it in public meetings and they do it when they are hired to work on these projects. You can't get a penguin to do that."

Whenever important visitors came to town, Glenn toured them around the weather-beaten compound and showed BASC's plan for a new building. He used education funding to bring in top scientists for lectures and took them on tours, too. BASC produced a history of NARL and published proceedings of two scientific workshops to document the need for new facilities. It was hard to disagree when Glenn, powered by his natural righteousness, laid out this homework and showed the sorry state of the buildings. But that was only backup for Richard's work. Richard knew how to get information discreetly into the hands of decision makers from the people they trusted—the kind of communication that leads to action in politics. When Oliver sent him to Washington to work getting ANWR into the energy bill in April of 2002, he took the opportunity. Nothing was happening—Alaska's senators stalled ANWR because they didn't have the votes—so Richard worked on funding for the new BASC building.

Ted Stevens already cared about Arctic climate change and research. He had seen with his own eyes and heard from Native friends how dramatically the environment had changed in Alaska, especially on the Arctic coast. While he was skeptical about the cause or worldwide nature of the change, he believed it required action. Ted's old friend Senator Robert Byrd, the venerable West Virginia Democrat, approached him with a bill calling for a crash research program for technology to address climate change, to be coordinated by a new climate czar in the White House—an unwelcome idea to the Bush administration, which was doing its best to pretend the issue didn't exist. Against staff advice, Ted cosponsored the Byrd-Stevens Climate Change Strategy and Technology Innovation Act and had it inserted into the Senate's energy bill.

Ted was a loyal old Republican, but politically he was invulnerable and could do essentially whatever he liked. In Alaska, he stood above mortal politicians like an omnipotent Olympian god. Starting with his important role in the Eisenhower Interior Department bringing about Alaska's statehood, and continuing as a senator since 1968, Ted had influenced almost every aspect of Alaska's modern political, economic, and social life and was arguably the single most important figure in its political history, living or dead. Besides that, the money he brought home, while making him a notorious porker in Washington, D.C., made him, personally, one of the largest segments of the Alaska economy. Nothing Ted Stevens could do would imperil his reelection for as long as he wanted to stay in Washington. In his last two elections the Democrats had showed so little interest they nominated candidates many considered mentally unstable.

Richard's uncle, Borough Mayor George Ahmaogak, contacted Ted about the need for the building. Richard and Glenn worked with Syun-Ichi Akasofu, a NARL alumnus, noted in scientific circles for helping establish the modern understanding of the aurora based on data gathered in the International Geophysical Year. He was a master science administrator and political operator who had won large special appropriations before. Ted trusted Syun to give him good advice about science and they talked frequently. Richard also pushed Oliver to use his friendship and influence with Ted. Oliver said, "Richard has been bothering me all winter, giving me big briefing papers. I said, 'Just give me some bullets, I'll bullshit my way through it.' " He made the call and Ted said he would get money for the building.

Glenn Sheehan popped out of his office at BASC waving a faxed press release in the air: $35 million for a new Barrow Arctic Research Center had been included in the Senate's version of the energy bill.

People appeared and disappeared constantly from the NARL Hotel, the Ilisagvik cafeteria, the ARF and BASC labs. Although the researchers were collegial and even intimate, as people often become when they are thrown together in a strange, remote place, a sense of community didn't have a chance to develop. Instead, their groups intersected in the cafeteria like the overlapping slivers of circles in a Venn diagram. You could look around the room and pick out which tables represented which projects: the Aerosonde guys, the snow transect team, the college students, the maintenance guys, the people from Wildlife. Scientists reconnected with colleagues they had met in Antarctica or through common acquaintance with certain teachers or well-known PIs. I sat at different tables and answered questions about the Natives' whaling for the scientists and about the scientists' research for the Natives. Glenn Sheehan and Anne Jensen table-hopped, too, but even their circle had limits. When Oliver and Annie Leavitt came in for lunch one day and sat with Glenn and Anne, I found, to my amazement, that I needed to make introductions and break the ice with stories of my misadventures as an inexperienced Arctic snowmachiner.

Jim Maslanik was in the cafeteria, looking as happy and confident as at any time I had seen him. He was going home to his family soon. He had a year ahead of him to plan the satellite validation project, using the knowledge gained in the field with Matthew. Next year he planned to hire someone like Warren Matumeak again. Warren had taught him less than he hoped about snow and ice, but it felt much safer traveling under the protection of an elder's knowledge of ice conditions and navigation. Jim still had work to do pinning down the precise locations on the surface chosen from the satellite and on coordinating measurements from the different levels in the air above the same spot. He wanted to include a manned small plane in the project, not just the Aerosonde, so that he and the plane could see each other and line up. The NASA P-3 would fly above the smaller plane. He also was happy with the Aerosonde, another project he was contributing to. It had set a new record for continuous operations while working with the Department of Energy, forty-eight hours straight. When the Aerosonde guys in Barrow went to bed, they had turned control over to co-workers in Melbourne, who flew from a desktop computer. Aerosonde controllers could communicate with the plane's radio connection through the Internet or, beyond its hundred-mile radio range, make the link to its onboard modem by calling it up through an Iridium satellite telephone.

The Aerosonde guys would be back in the fall for more tests and operation. Jim Maslanik would be back with them. Matthew Sturm came back all the time. But April Cheuvront knew she might never be back.

April's circle in the cafeteria reached only as far as the snow transect team, and it was evaporating. She missed her husband, Steve, her students, and her little cabin in the mountains of North Carolina, where spring was already well advanced with blooming azaleas and rhododendron. But she also felt trepidation after Matthew's warnings and felt loss, with her comrades dispersing and the ensemble relationship of the transect disconnecting into the impersonal setting of the NARL Hotel and the cafeteria. She looked a little nervous sitting at a table with trays and dishes beside Matthew and Ken, laughing but unsure of herself, with another emotion just below the surface.

She slung her duffel into the back of a borrowed pickup and drove with Matthew to the middle school to speak to a class of sixth graders before going to the airport. For all his brilliance, Matthew had trouble finding the kids' level and started his talk by trying to define the word environment—"What do we mean by a big word like that?"—while they were more interested in the types of snow that provided optimal habitat for different kinds of animals. April revived their attention by talking about home. "Baseball season is about closing up now. I missed baseball season," she said. "Roads go everywhere. There's not much wild, open land, which is what I love in Alaska. I live in the mountains about twenty-five miles from my school, and most of my students have never been that far from home." A show of hands demonstrated that all fourteen kids in the class had been at least to Seattle, most farther.

The one-room airport was still empty, well in advance of April's flight. She put her arms around Matthew's neck and said, "I promised myself I wasn't going to cry," her eyes overflowing. Matthew said, "I feel like you're my daughter." April slipped some folded notebook paper to him, letters to the two team members she didn't get to say good-bye to before they left. Matthew drove off in the pickup on an errand to borrow emergency locator beacons from Search and Rescue for use on the ice project with Jim Maslanik. April sat alone in the airport and sobbed. Later that day Matthew got a call from a colleague who ran into her in the Ted Stevens Anchorage International Airport while she was waiting to change planes, reporting, "She isn't decompressing well."

Matthew fulfilled his obligation to report back to the Barrow community with a well-attended evening talk and slide show organized by BASC at the Heritage Center. He was an engaging speaker: "It was fantastic to see Alaska unroll from Nome to Barrow. I haven't really processed it. The trip is just a few days old, but I'll never forget it. I've lived in Alaska most of my life, and seen a lot of it, but I never knew how beautiful it is, or how much there is where absolutely no one ever goes." The audience wanted to know about the scientific and the practical aspects of the trip—what gear worked best for such rugged use—and how April handled it, as if her durability were also part of the experiment. Matthew said she was a good hand and praised her openness and ability to engage students in each village along the transect. But if he included a teacher again, he said, "I'd be even more careful. You don't realize, if you displace someone that far, will they survive. When I saw her off at the airport, I sort of realized how hard a thing I was asking her to do. It's tough."

Matthew explained to me later what he meant: it wasn't just that April was headed home to find out what life was like without her. She also was headed away from the Arctic, possibly for good, while the rest of the team would all be back. "For us there's always another expedition," he said. "For her, there's just another year in the classroom. This had to be a high point in her life." After an intense love affair with an utterly new and unfamiliar place and kind of life, she would be severed from the north and placed back into the unbearably mundane south. Later April agreed—she nodded emphatically—yes, it was a love affair, ended. Which was exactly how she looked sitting in the airport in Barrow as she prepared to fly home.

The colors blasted her first, all the green trees and vegetation in an overwhelming swirl. The jet lag messed her up, too. Steve stood waiting when she got off the plane in Charlotte. He had missed her. But it was herself April had to worry about. She was freaking out over a bridesmaid's dress. All the way home, up the mountain, she worried frantically that her new upper-body muscle wouldn't fit the dress for her brother's wedding. After stepping on the scale and discovering a gain of fourteen pounds, she made Steve drive her right back down the mountain, thirty miles to the dress shop. The dress could be altered—it wasn't really such a big deal—but April couldn't help overreacting. It seemed an omen of how poorly she would fit back into her life.

When she returned to school, April found it was just as Matthew had predicted—life had gone on, students had forgotten about her. Worse, her co-workers resented her. She had become a minor celebrity, with her journal serialized in the local newspaper, and awarded Burke County Teacher of the Year in her absence. Other teachers made it clear they thought a twenty-five-year veteran should have won, not her. In her classroom, it was like being a new teacher all over again. One of the kids said, "You're the sub for the substitute."

After a week, the truth slipped out. About halfway through the transect, the substitute had abandoned the lessons April was sending back, the lessons she had stayed up into the morning to produce with Glen Liston after long days of hard work on the snow, and reverted to the textbook instead. April was crushed and humiliated.

The one person who understood what she was going through, April said, was her father, the old firefighter, who told her the story of his return from Vietnam. She said her experience couldn't compare, but he insisted it was the same. He had been drafted and then reenlisted for a second tour. His family stopped writing; connections with home disappeared. When he got home, it was as if he had never been born; there was no place left for him. After a month or two he got ready to leave forever, to go to Thailand to run heavy equipment, when he met April's mother, a teacher and a strong Catholic, who settled him down back home in West Virginia.

In Alaska, they were still betting on how long it would take April to break away from Morganton. Another teacher in the same program had moved to a tiny Alaska village. At twenty-six, April would have to make a decision soon. She and Steve had bought a piece of land up in the mountains to build a new home.

CHAPTER FIVE

The Ice

THE AHMAOGAK HOUSE was full of family, even more than usual. A rough count of sixteen seemed reasonable, in the little dining room and the attached living room, but keeping exact track wasn't possible, as teenagers and adults entered constantly and peeled off heavy parkas, receiving loud greetings, or rewrapped themselves to depart for short errands, going for food or some item that was needed. Counting the swarm of black-haired toddlers was hopeless. Richard Glenn arrived with his wife, Arlene, whose seven-months-pregnant shape bulged under a lovely fox-trimmed parka of her own making, and greeted his uncle Savik and aunt Myrna Ahmaogak. Richard was back from Washington just in time to help Savik Crew with final preparations for spring whaling.

While gathering supplies and equipment, family members also celebrated Savik and Myrna Ahmaogak's anniversary (they wouldn't say which one). Savik was the brother of Richard's mother, Kannik; Mayor George Ahmaogak was another brother. The table was full of food and surrounded by people eating, and Savik sat back with a big grin and surveyed all the activity through his thick glasses with a patriarch's satisfaction. Myrna worked efficiently, moving between her place at the table, her little kitchen, and the ever-brewing coffee machine—a big metal Bunn machine with two pots, like those found in restaurants. Her hands moved fast over the maktak with a sharp *ulu,* the semicircular knife that's customary in Iñupiaq food preparation, and her authoritative voice and expressive face managed the flow of generations, steering the hungry into chairs as soon as they became available so everyone got a turn to eat.

Richard and Arlene had just fed me a big meal, but I couldn't have resisted Myrna if I had wanted to and took a chair still warm from a son-in-law who had moved over to the other side of the room, near the TV and harpoons. A big cauldron of caribou stew stood in the middle of the table, a flavorful broth with rice, potatoes, and hunks of meat still attached to the bone. Each diner ladled out a bowl of soup with one big meaty piece, then cut off bite-sized chunks to dip one at a time in the pot of seal oil, which added a rich, dusky flavor, something like sage but with as many complex undertones as a fine wine. The glistening maktak traveled around the table on a cutting board along with an organ I didn't recognize that looked something like a bent vacuum cleaner hose. The cutting board stopped long enough for each person to cut and eat his or her fill, with a few nibbles left chopped off a big pink block of blubber to get the next person started.

Glass covered the oval top of the wooden table, and under the glass Savik and Myrna displayed photos of grandchildren, whales, and boats and sayings, flags, maps, and documents for quick reference, such as a list of Barrow whaling crews and captains. Every wall was covered with family pictures, too, and whaling pictures, prayers and religious posters, as well as patriotic items and a September 11 memorial. Savik showed off some highlights among these, especially a picture of the whole family lined up in front of an enormous whale the crew had landed several years earlier, the only picture in which the group looked small.

The Birthday Show was on, the KBRW chorus of best wishes, coming through the cable TV Eskimo Channel, a broadcast of town notices and happenings. There were many wishes for Savik and Myrna's happy anniversary. Someone handed the phone to Richard and he added his own congratulations to the stream with a smile of pleasure, in perfect time with the rhythm of talk. Then he shooed some rambunctious children away from the harpoons and the bomb-shooting shoulder gun, observing that as long as he could remember, this house had always had a flock of children under five years old tumbling over one another—was that possible?

While the elders were smiling and considering that question, a big, store-bought sheet cake materialized and pieces thick with white frosting began circulating along with mugs of hot coffee.

Richard and his cousin Roy Ahmaogak, Savik and Myrna's son, sat on the sofas facing the TV amid the activity, inspecting the weapons. Each harpoon had a brass darting gun attached, a simple but ingenious Yankee whaling invention that on contact with a whale's body would fire into it a pointed, foot-long brass bomb with a fuse of five to eight seconds. The traditional bomb was a brass tube loaded with black powder and a fuse; the newer, stainless steel superbomb, introduced to kill whales more quickly and humanely, used more powerful high-tech explosives. The shoulder gun also shot these bombs, with a legendary kick, from a brass barrel that looked like a cross between an old-fashioned shotgun and a cannon. Either way, in the darting gun or the shoulder gun, the bomb was propelled by black powder hand-packed with wadding in a reusable brass shell casing. Richard and Roy had plenty of these shells left over in the bomb box, a big, sturdy plastic tool box also full of bombs, black powder, and tools. Roy thought the shells were probably good for another year. Richard preferred to repack all of them.

Roy and Richard had come up as whalers like brothers, and they still related that way. Savik was the captain, but he rarely went out to camp anymore and left most of the decisions and judgment calls to the younger men, who functioned as cocaptains. Roy, whose only occupation was subsistence, was usually at his parents' house when I came by; he lived in an adjoining apartment. Out of Richard's presence he talked about guiding him, helping him develop a whaler's skills, giving him responsibility—although Roy was only a few years older and no more experienced in whaling. Richard was harder to find, as he was among the most important leaders in town, an ASRC executive, a political operative, and the president of BASC, and was busy with Arlene and his daughters. In the boat he usually took charge. But although Richard bore Savik's name as his own Eskimo name—Little Savik, they called him sometimes—Roy was Savik's real son.

In 1959, Richard's father, Bob Glenn, was working at the air force radar station near Barrow—part of the Cold War system that scanned for transpolar attack—when he became interested in Kannik, who worked for an airline, through daily radio exchanges of weather information. As soon as they met in person, Bob's employers forbade any further contact with the village. He had to quit his job to marry a Native girl. They moved to Sunnyvale, California—Silicon Valley—for work, where Richard and his two siblings were born. "You get so homesick you want to die," Kannik recalled. "I was so scared, so afraid of people."

Like other immigrants swirling in the melting pot, Kannik fought to keep the old country alive in her children, feeding them Eskimo food her mother sent from Barrow, speaking Iñupiaq in their hearing, and exchanging tapes with family. Several Iñupiaq families lived in the neighboring towns and more attended vocational school nearby. They joined for celebrations and funerals and hosted friends and relations who passed through from home, salving their homesickness with maktak and Eskimo dance. Oliver Leavitt, a few years younger than Kannik, lived in the Bay Area then and was sent by his mother to check on her and the children when Bob was away in Greenland for seventeen months. Richard, around age four, idolized him. Oliver would say, "Look handsome, Richard," and Richard would raise his right hand in a salute. Richard's quick mind and ability to read people was clear from a young age. His father helped develop his talent in mathematics. Kannik worked gently to plant a love of his other home. Richard grew with the potential of two worlds within him and a latent choice of which one to make his own.

The choice became concrete around age thirteen, when Richard traveled to Barrow for the summer and started learning to hunt with Roy, Savik, and grandfather Walton Ipalook Ahmaogak. Savik and Myrna's tiny house was already crammed with eight children, so Richard set up a tent by the front door. He learned to follow animals, navigate on the tundra, coil a rope, butcher a kill. He learned the traditional way, by doing, getting it wrong, adding his own personal contribution to centuries of Iñupiaq trial and error, and thereby deriving a feeling for how nature worked. In error, he also learned the value of humility: that unless you could admit you were not ready to go, you were left behind; that pride was dangerous when it kept you from admitting you were cold. It took courage to attempt what a teacher assumed you could do and strength to withstand the consequences of failure—the ego-grinding disapproval and deflating humor. But Richard did learn and he gloried in it, gloried in the freedom and the belonging his new knowledge afforded. In the fall he prayed for fog to keep away the airplane that would return him to the mundane world of a California high school. "Here's a whole nation of relatives, everything north of the Arctic Circle, and here is a land with no highways, no fences. Who wouldn't want to be a part of it?" he said. "I started throwing newspapers, doing everything I could to come back here every summer.

"You wonder what inside you is worth keeping and what is just shopping malls and gas stations."

For Kannik, Richard's decision to take his place in Barrow was the joy of her life. For Richard, it became a mission. While he excelled in school in California—math, science, and debate—he thought of himself as an Eskimo and his home as the Arctic. This amphibious ability to thrive above or below the surface of the cultural divide came to define his life. Richard's gift and burden—in business, politics, and his relationship with Roy—was to see on many levels. A gift because it helped make him a master of two cultures, a friend to everyone, able to navigate through extraordinary conflict. A burden because, to Richard, every mistake told a story about personality, psychology, and character, either his own or someone else's. But sometimes a mistake is just a mistake.

The first time I met Richard he told me a story about himself and Roy and a mistake. It was 1997, during the spring whale hunt. A dozen crews were camped on the ice about seven miles out with a whale coming in to be butchered. The water was glassy calm, but at the ice edge chunks began tumbling upward to the surface. Roy said, "Grandfather always said, when the ice is like that, you should go home." Richard evaluated the situation. He saw older whalers staying, men who had seen everything on the ice. One of them told him the trail home looked fine. "I said, 'I think we should stay,' " Richard recalled. "I put my foot down. And right about the time I put my foot down was when the ice broke and we started heading out to sea at about two knots."

A fast-growing new crack sheared off a floe more than forty miles long. Once loose, that ice sheet broke into smaller pieces. The newly open leads quickly produced an impenetrable fog. More than fifty crew members were able to rush across to safety in time—one woman jumped her snowmachine over a moving gap four feet wide—but within hours 154 whalers were stranded on dwindling fragments. Ninety were crowded on one piece as small as a football field.

A break-off with people on it has its own word—*uisauniq*—so it is not a new occurrence. Sometimes, in the old days, hunters simply disappeared, floating off to oblivion. Other times, they were lucky enough to float along shore, hiking and boating to maintain their position as the ice rotated, getting off one floe to board another and work closer to the beach, and ultimately finding a way to land after days and many miles of drifting. But break-offs happened infrequently, and never with so many people. Otherwise, the villages might have disappeared, as in the old days when the populations of Utqiagvik and Nuvuk were counted in the hundreds; a couple of break-offs like the one in 1997 would have wiped them out.

But no one died in 1997, thanks to technology. The courageous pilots of the North Slope Borough's Search and Rescue helicopters flew to coordinates whalers read off their GPS receivers, descending blind into the fog with guidance over VHF radio from people below who could hear their engines in the whiteness. Some crews burned clothing to make beacons for the choppers. Pilots landed on shrinking fragments of ice, taking women and children first while the men stood guard against prowling polar bears. With each trip back and forth, the ice moved farther out to sea and broke into smaller pieces. Some whalers made it home in boats, following directions radioed to them by the helicopters above. Before the last of the gear was rescued, the floe had moved thirty-five miles off shore.

Was climate change responsible for the break-off? Richard was impatient with the question. How could anyone say? To him, this was a story about humility. One of the strongest lessons of Iñupiaq culture was that to make good judgments about nature one must be humble. The information he had needed to make the right decision had been in front of him. Roy had seen it. But Richard, in his own interpretation, had been too strong-willed. He had let his pride blind him; he had put his foot down.

When I met Richard and Roy together for the first time, Richard told me the story again. A humble bow to Roy. And he told it other times, as if to remind himself. Roy, he freely admitted, despite the competition and "head games," knew a lot, more than he did even.

On the subject of the shells for the darting gun and shoulder gun, however, Richard's point of view represented caution. Damp powder in just one shell could mean losing a whale. Somehow, without open disagreement or confrontation—these negotiations were always too subtle and low-voiced for me to follow—Richard and Roy arrived at a compromise.

They would test the shells. A perfect solution: whether or not the shells went off, they would have to be repacked, as Richard preferred, but if none failed Roy would still be proved right.

Richard, Roy, Arlene, and two teenagers, Benny and Eben, rode in the pickup to a gravel pit with the shoulder gun and the questioned shells. Eben, a big boy, was the new harpooner for this season, although without a whale yet to his name, and Benny was learning harpooning. Savik's plan was to give each of the teens an opportunity to strike a whale, bringing both up to greater responsibility. One loaded up a shell (without a bomb, of course) and latched shut the shoulder gun. It looked to me a dubious contraption that belonged in a museum, and I couldn't help stepping back—which didn't go unnoticed, although no one said anything. The first shell went off, spraying wadding with a boom. The next went off. Several more worked before one failed. Roy told the boys to open the gun and rotate that shell, but it still wouldn't fire. It was a dud. Richard said, "You see, that's why we have to repack every year."

The men retreated to Roy's apartment to repack the shells. Richard and Roy set out the tools, shells, and explosive black powder, working carefully with quiet conversation. Richard said he had been taught to use seventy grains of black powder in each shell; he had assumed that was approximate, but the first time he put in a little extra the bomb shot right through the whale and came out on the other side, exploding under water. It was a time to tell such stories, to catch up after weeks apart. Richard, a master of mimicry, kept everyone chuckling about the lobbyists and congressional aides he had spent weeks with in Washington. He and an Iñupiaq colleague had given each of the white men an appropriately ridiculous Eskimo name—words meaning things like dimwit or brownnoser—but with nearly accurate translations that made the names sound impressive so that their unsuspecting victims would go forth proudly repeating them. I asked what they had named Ted Stevens, but Richard shook his head: they would never do such a thing. Roy suggested Umialik as Ted's name. Richard said he probably already had at least one Eskimo name.

For the Iñupiat, names matter. According to tradition, the spirits of ancestors wait to enter children who receive their names. Fretful babies have been known to calm down when named, or even to know things that only the deceased elder would have known. Some are named right away, others later; some get one name, others are named many times by various elders through the course of their lives, as the older generation perceives something in the younger that brings back the memory of a face that is gone. I suppose the tradition explains the still-common use of nineteenth-century English names such as Ambrose and Flossie, as well as the survival of Iñupiaq names.

Roy started casting about for a name for me, which evidently made Richard nervous. Finally he suggested Malik. He said it meant follower, since I followed people around asking them questions. I thought I was probably the butt of another joke, but Richard quickly added that the name belonged to one of Barrow's greatest whaling elders and referred to his unerring ability to follow the whale to where it would surface. Malik had caught many a whale as a harpooner for Savik Crew and for many other crews in Barrow and had taught many whalers, including Roy—probably more than any other elder alive. He was known for his skill and courage in killing whales the old-fashioned way. Instead of firing additional bombs into a whale, Malik would climb out of the boat onto the whale's back and deliver the fatal thrust with the lance as it swam. The name obviously was unearned by me. Roy was the only one ever to use it, and he only a few times. When I mentioned it to Oliver later he just grunted and turned away.

When the weapon work was done and we had returned to the quieted house, Savik put a videotape on the big TV while Richard showed Eben how to oil the darting gun, propping it on the coffee table in front of the screen. The images were jerky and washed out—women in a pickup truck wearing parkas with fur ruffs, the gray ocean outside—the sound was of cheers and laughter, voices shouting gladly over the radio. This was a family video of one of Savik Crew's first whales, from a fall season in the late 1980s, not long after George and Savik Ahmaogak split their crew in two. A man ran along the gravel beach, holding aloft in one hand a flag with a knife on it. Richard sat back and narrated. It was Savik Crew's flag—*savik* means knife, the straight-handled hunting kind, not the round ulu. A crewman must run to carry news of a whale with the crew's flag held high, the only time the crew is entitled to fly the flag, and place it on the highest point on the captain's house. The camera followed him in profile from the truck with the flag held high—how could he run for so long, so fast, except with the power of great joy? Now the same house in which we were sitting, young men climbing higher and higher on the radio antenna, carrying the flag far aloft. Then the serving, a great crowd of people coming forward for big platters of food set in front of the house on metal drums—maktak, kidneys, intestines, flippers, baked bread. Those seated with me on the couch called out the names of people they saw. Myrna serving, Savik smiling, Richard, with all his hair in those days, and Thomas Itta, Sr., as a younger man—he was a member of Savik Crew in those days, too. And the flag flying above it all.

Richard spoke clearly, deeply to the room: "That's what everyone thinks about when they are falling asleep, the day when the flag will fly over their house."

The east wind let up on the evening of May 1, 2002, and Richard and Roy went out on the ice to scout trail. Warren Matumeak and some others believed the ice should be relatively good: the west wind had consolidated it against the point and then the east wind had blown away the poorly bonded pieces. Richard's analysis was more cautious. The ice was still new, with few anchors, and he had been taught that a day or two after a big east wind can be a dangerous time for break-offs. The scientific theory behind this was that the east wind would push water away from shore but that, after the storm ended, the water would come sloshing back, raising up the ice and cracking it while pulling it directly away from shore with a surface current. The 1997 break-off occurred under similar circumstances: the ice was weak and young because of a March break-off, there was a lot of open water, and there had been a strong east wind earlier in the month.

Roy led our group of three on snowmachines down the beach to the south of town. Before the storm, while Richard was in Washington, Roy and the boys had built a trail and hauled the umiaq and aluminum boat out here, then rescued the boats as the wind moved the ice relative to the shore. Normally Savik Crew went north with George Ahmaogak's crew, up near Nuvuk, where ABC Crew was camped. The other alternative was the middle section, off NAPA, where Oliver Leavitt Crew was helping build a trail at an ice embayment that brought the edge closer to town. Roy's idea was to catch a whale away from all of them, at the south end, farther from shore. Richard didn't like the idea, but he didn't say anything definite against it.

Roy's route twisted back and forth to make the most of flat floes, disappearing behind low ridges. He stopped at a sizable watery crack, where he and Richard checked the current by lowering an old spark plug on a string. The going wasn't too rough until we reached a ridge as long and high as a river levee. We turned off the machines, climbed up, and sat looking for a long time at impassable broken ice. The orange evening sun backlit shattered shapes of light blue and darker blue, an endless wreck, extending as far as the horizon, like a silent, devastated kingdom. We were five miles southwest of Barrow, and Richard said there were at least five miles of bad ice ahead. He stared long at the horizon. Roy talked casually. He joked—only half joked, I thought—that he wanted to get home for *Seinfeld* at 10:30.

Back home, Richard's beautiful eldest daughter, Patuk, was writing out graduation announcements, pushing an impossibly heavy pen, a study in youthful languor. She had a little over a week to pass her Algebra II class but laughed at her father's fear she might not make it. Everything was so easy for Patuk, Richard was afraid she was in for a rude surprise someday. Alice, the thirteen-year-old, detailed the cracks of her immaculate sneakers with a bleach rag. 'Berta was asleep; at age nine she either played energetically outside or curled up to sleep with her beloved dog. Arlene, disabled by her pregnant belly, presided from a rocking chair. Everyone was laughing and teasing; Richard didn't stand a chance. He hung his head low, shaking it in defeat as if hinged at the shoulders, his eyes creased shut by his smile.

Max Brewer, the NARL director called King of the Arctic, said Arlene might have been older if he hadn't sent her father, Frankie Akpik, to manage a drifting ice station for five years before her birth. As each of his girls was born, Frankie would pass out cigars in the NARL labs, saying, "Jeanette got another helper." The scientists knew young Arlene as "The Princess." She was regal like a princess, quiet and dignified but with a smart, mischievous smile that suggested she knew a secret. Even puffed up with a fourth baby, nearing forty, she retained that girlish, dimpled smile, along with a quick intelligence and a focused will that were more than a match for Richard.

Richard and Arlene had paired up when they were teens Patuk's age. She picked him out one day, a handsome but oblivious boy, as he walked by outside the window of the Presbyterian Church, where she was working. She had heard his name on the birthday program that morning and leaned out to wish him a happy birthday. For two years, while Richard was away at college in California, the relationship survived long distance; for Richard, being faithful to Arlene also meant being faithful to Barrow and the life he coveted there. They were married at twenty and had Patuk right away, moving to the University of Alaska Fairbanks, where they lived in an old log cabin down a dirt road near a slough, just off campus but seemingly deep in the woods. Arlene sought a degree in Iñupiaq and Richard learned Iñupiaq and German—he has a facility for languages. He studied geology because it put him near the land, completing his master's degree with a thesis about rocks on the northern side of the Brooks Range. Arlene went to work for the Iñupiat Commission on History, Language, and Culture, recording and preserving her people's tradition with audiotapes, artifact collections, and carefully translated volumes of elders' conferences. From the tools of academia and modernity, they built a traditional Eskimo life and family together.

For the modern Iñupiat's brightest, Richard and Arlene's path of departure and return wasn't unusual. Traditionally, there was no adolescence in their culture. The village had too much work to do and life was too short to include a decade of self-exploration. In the nineteenth century an elder might be thirty years old and sixty was extreme old age. In the early twentieth century, Kenny Toovak and Thomas Itta, Sr., each spent only a few years in school because their families needed them to hunt. But those born in the latter half of the twentieth century went through high school and learned its social lessons of independence and rebellion, the message permeating the pop culture of breaking away from parents and overturning expectations, of self-definition through the talismans of brand marketing and adherence to the musical tastes assigned to each age group. When that phase of self-definition ends, most people end up with culturally defined roles—as parents, employees, church members, citizens of a community. People attach somewhere, either by design, in a place and with family and a career they have chosen, or randomly, like seeds rooting where the wind leaves them. For the Iñupiat, however, the day of attaching can arrive with a disabling cultural deficit: young people who haven't learned their language or the skills needed to connect with traditional knowledge may not be ready to contribute to their community as full adults. I met several who had graduated from college and returned to Barrow ready to take their place—they had experienced and rejected the alternative "mall culture"—and now began for the first time sincerely to learn to speak, to hunt and prepare food, and to know the old stories.

Those who master both worlds, as Richard and Arlene did, could become treasures to their people. Arlene used the skills of a linguist and collector to understand her own culture and act as a bridge to outsiders and to the future. Richard used the skills of a geologist to develop a natural gas field sufficient to supply Barrow's energy needs for generations. "My kids and grandchildren will be warm," he said.

They lived in a square little house on the edge of Barrow the size of a low-cost apartment in a typical fourplex in Anchorage but, by Barrow standards, modern and light. Richard enjoyed having his family close at hand. His pride was his ice cellar out back. Under a half sheet of plywood a ladder descended sixteen feet into the ground to a big round room where, after work each day, he enjoyed swinging a pickax against the frozen earth, extending walls a fraction of an inch at a blow. The governor had been down there for a tour. Now it was clean, a new layer of snow on the floor, waiting for whale.

Back in the living room Richard checked the National Weather Service Web site for satellite pictures of the sea ice and the forecast from the ice desk. He had a lot of respect for the lone ice forecaster in Anchorage, Russell Page, a former air force guy from South Carolina who had set out to develop the skill of predicting shore ice events at the request of Maggie Ahmaogak some five years earlier. Page's forecast, which came up as a hand-drawn map covering most of Alaska, predicted break-off conditions around Barrow.

The next evening, Richard went out on his own to chip trail. He wanted to go north of the embayment where Oliver and the others were setting up camp, but south of Uncle George and the group off Nuvuk. He thought he had a section of ice with an east-west edge where whales might emerge. Driving his snowmachine over steep ice and chipping and filling where driving was impossible, he pushed by midnight to within sight of the lead about three miles off the NARL runway. Roy and a cousin from Wainwright named Duane joined him the next day after lunch under blue sky and warm sun. Roy still wanted to go to the south. On Richard's rough trail I soon dumped my machine in a hole where I couldn't get it out without help to dig and push. Roy joked, "This isn't a trail, it's an obscenity." Using the technique of standing on high points at opposite ends of the trail and then approaching each other over the smoothest areas available, Richard and Roy found a better route that was passable after an hour or two of chipping the rough spots. The work placed us back at that same long, high ridge that looked like a levee, and beyond it the way appeared crazily rough, just a jumble of broken ice similar to the area that had blocked Roy's trail, although not as wide. They decided to walk to the edge.

We were like ants crossing a rock pile, climbing up and down over a texture whose scale was larger than we were. Junk ice, as Richard said. Before the west wind's storm had piled up these ridges, just a few days earlier, this had been flat new ice floating somewhere out in the ocean. Back then a light snow fell and polar bears and foxes left crisp tracks on the level plain. Those tracks still remained clear and well defined, but now they ran vertical, traversing impossible angles. A fury of collisions had turned flat ice into a landscape of little mountains. Richard and Roy examined it with two questions in mind: would the new conglomeration hold without breaking off, and could they find a flat place strong enough and large enough to pull up a whale? If yes, then it might be worthwhile for the crew to chip its way through the ridges to the edge, given enough time and enough cookies, coffee, and maktak.

At the edge, the ocean never looked so liquid, its waves slapping the ice, its color the darkest blue-green, almost black, but flashing on the surface with the brilliance of the sun. There were flat places, small ice floes that had docked but had not been compressed and broken by further collisions. We walked along the edge from one flat floe to the next until we came to a lookout tower of ice, an exclamation point that rose right above the water, and there the men climbed up and commenced their inscrutable Eskimo stillness, gazing in silence at their surroundings.

I was glad for the rest. I put down my heavy shoulder bag. It contained my own handheld VHF radio, GPS receiver, and laptop computer, and all the yellow notebooks I had filled, so precious I had promised myself never to be parted from them until they were safe at home in my black filing cabinet. Richard and Roy had already teased me about the bag, and I had admitted I was paranoid. I watched the dark waves rolling in and listened to the splash and rumble as they hit the ice edge and slid underneath. It felt like we were moving, but I assumed that was an illusion. I lined up the horizon with a mark on the ice. It did appear we were moving slightly, but that was probably just my own breath and heartbeat. I braced myself against vertical ice and looked again. I said, "Richard, I think we're moving a little bit."

Richard, seated up above me, said nothing for a moment, just kept looking. Then there was a horrible groaning crunch and the whole tower where we were seated started to move. Richard yelled something and the three Iñupiat moved lightning fast off the tower, over a low ridge, and across the nearest flat ice pan. Adrenaline shot up my spine to my scalp like an electric shock as I scrambled after them, part of my imagination already visualizing a fall into the frigid water from an overturning iceberg. My bag swung around my neck and threw me off balance, bringing me down on a shoulder. The others were already well ahead. I rose, still tangled with the bag, which pulled me sideways like a drunk while I tried to regain my balance over the rough ice. In seconds it was over and I was standing beyond the flat ice with Richard, Roy, and Duane, all laughing together to expel the remnants of that clammy jolt of fear.

Myrna had another feast on the table and lots of family gathered around when we got home. Over the VHF the news arrived that ABC Crew had landed a whale. Cries of "Hey-hey-hey," congratulations for the first whale of the year. Crews prepared to send members up north to ABC Camp, off the point, to help pull it in and butcher it. I joined Eben and Benny, riding along the coast and then out into the pressure ridges on a confusing trail that wove its way to the ice edge. Dozens of people were there in the gathering dusk chipping at a high ridge to make a runway where the whale could be brought up. It was late when the whale arrived, pulled behind the small aluminum boat that had killed it, with the crew's flag flying over the stern. A small whale, thirty-eight feet—its flesh would be tender, but still it would take all night to butcher. Polar bears soon became a problem. Late into the night, as tired workers cut, a bear came within a few feet of one of them, unobserved, and had to be killed.

That night it was decided that Savik Crew would go to George's spot off Nuvuk, north of the ABC Camp, and whale there with the aluminum boat, too. Richard and Roy's trail experiments were abandoned. Richard led a crew of six, but Roy stayed behind (he told me he was giving Richard a chance to lead; Richard said Roy might be disgruntled about not going south to his preferred trail). By around 3:00 a.m. the crew had established camp and the night shift was at the ice edge with the boat, watching for whales.

A call in Iñupiaq came over VHF channel 72. A big piece of ice had broken free ten miles up the lead and was spinning in the water and headed north. The caller was pulling back. Where Richard was standing the current was going the other way and conditions looked good, but he didn't engage in a debate. He came back to the other captain with a statement that translated as: "Whatever you do, we are going to do the same." Twenty crews up and down the lead listened in silence for the answer. The other whaling captain stood by his assessment; that dangerous spinning ice floe was headed their way and he was pulling back. Richard woke the crew, who had just gone to sleep after a long day of hard work. They broke camp and pulled back a couple of miles to a rounded multiyear ice floe nearer shore, then set everything up again and bedded down as day started. Richard said, "Your eyes are beyond your own because you're listening to somebody ten or twenty miles away and you're trusting what they tell you."

Sea ice is an intrinsically fascinating and enigmatic material. Put a glass of salt water in the freezer, as my daughter Julia did for her first-grade science fair project (two and a half teaspoons of table salt in a twelve-ounce glass of water is about right to get ocean salinity of roughly 3.5 percent). Unlike the clear ice freshwater makes, saltwater creates exquisite crystals that look like flat plates, randomly interlocking with gaps between them, which are especially visible at the bottom of the glass, where a layer of liquid brine remains, even at very cold temperatures. That's because saltwater doesn't really freeze. Water molecules have to push salt molecules out of the way in order to crystallize, extra work that explains why the freezing temperature for ocean water is 2 degrees C lower than for freshwater. The liquid that is left, with all the salt and a little water, resides in pockets between the ice crystals, too salty to freeze. Pull back to a larger scale, and the patterns of brine and ice enlarge, as liquid brine droplets connect and carve downward-flowing channels through the ice, forming a precisely organized tracery of diagonal and vertical lines. You can partially re-create this in the kitchen, too, by turning your glass upside down and seeing the brine at the bottom melt its way through. Underneath sea ice, where the brine drips from the channels into the ocean, icy underwater filaments freeze around the supercold trickles (they were discovered only recently, using an unmanned submarine). The brine keeps going down—it is much heavier than the less salty water around it—sinking toward the colder, denser water layers near the bottom of the ocean. That motion helps power the upwelling of nutrients from the seafloor, which contributes to the richness of life at the ice edge. On the global scale, brine exclusion contributes to conveyer currents that transport heat northward on the ocean surface and move cold southward down below—a factor in climate. As the salt leaches out of the ice pack in the spring, the ice that remains becomes fresh. Now when it melts, a thin layer of freshwater spreads over the top of the saltwater in the leads.

The ice harbors life at the same range of scales. Tiny organisms live in microscopic brine pockets all through it, secreting slime that concentrates the salt to help keep their spaces open and in turn weakening the ice. Others can move through the ice, traveling vertically in brine channels and horizontally in layers whose conditions of freezing have created greater porosity. Plankton that later grow into larger organisms may fill either kind of gap. The multiyear ice also teems with small sea creatures that live nowhere else. Ice algae and tiny animals are food for ice-adapted fish and seafloor creatures, which in turn are food for whales, as well as birds, seals, and walrus, the last of which are food for the highest marine mammal predator, the polar bear. It lives almost entirely on the ice. In the old days, when the need arose to remove the flesh from bones such as polar bear skulls, the Iñupiat suspended them under the ice and let the unseen creatures there pick them clean—that's how abundant life is down below. At one remarkable site in the Bering Sea, eider ducks congregate in such great numbers that their body heat keeps open a window in the sea ice; they get their energy from the bounty of food that grows at the ice edge they help create. In 1993, a large creature like a sea worm washed up in Barrow that turned out to be a species new to science, a member of a phylum thought to be nearly absent except in fossils in the Burgess Shale. Who knows what else is down there?

The physics of the ice is strange and mysterious, too. What it most resembles is a high-speed version of the geology of the earth's crust, whose vast continental tectonic plates float and collide, fold and shatter into mountains. Pressure ridges often look exactly like miniature mountains. The sharp edge of an ice floe looks like the edge of the continental shelf. That similarity inspired the theory of continental drift, originally advanced by Alfred Wegener, a German polar scientist who died in Greenland in 1930 on an expedition of the Second International Polar Year. The inability to predict the behavior of these materials is another similarity. Indeed, the mechanics that build mountains and pressure ridges both remain too poorly understood to say if the analogy really holds on a deeper level.

One of America's greatest polar scientists, Norbert Untersteiner, whose measurements at an ice camp during the International Geophysical Year founded modern ice science in the West, spent much of his career trying to describe how the material behaved on a broad scale with a set of equations such as engineers use when calculating the strength or deformation of steel, wood, rock, and so on. He came to believe the task was impossible. "If you try to do that for the Arctic, you would spend the rest of your life calculating," he said. Besides the complex stresses and motion in every direction, the material itself is complex—it can be rigid or flexible, it can bounce back elastically or mold plastically. During the 1970s, Norbert sought to solve these problems four hundred miles north of Barrow on the most sophisticated ice expedition the United States ever mounted, the Arctic Ice Dynamics Joint Expedition, known as AIDJEX, which included 266 researchers and 900 tons of supplies (Kenny Toovak was a key support worker). The project set up three camps in a triangle a hundred kilometers on a side, with a fourth supporting camp in the center, all maintained for fourteen months. A key objective was to measure the stresses and deformation of the ice between the points of the triangle, but the ice cracked and the calculations were foiled. "It turned out essentially to be a flop," Norbert said. "We did not learn much at all. There is no universal dynamics we can concoct for this messy system."

Richard Glenn thought he might be the person to move the science forward when he set out to earn a Ph.D. studying the shore ice around Barrow in 1989. He believed he could combine within his one mind both Iñupiaq and scientific knowledge about ice, obtaining the benefit of both. He had a teacher at the University of Alaska Fairbanks, Lew Shapiro, who believed in him. They had met on a plane bound for different geology projects in the Brooks Range, a young whaler and a tough old former paratrooper from the Bronx, and found they had much in common. Lew knew Frankie Akpik from NARL and valued traditional knowledge; he considered Kenny Toovak a mentor and in the 1970s worked with him interviewing elders about their memories of ice events going back to the beginning of the century. Lew thought it should be possible to derive engineering equations for sea ice, at least under limited conditions, if you had a better understanding of the details of how ice was put together. A single sheet of sea ice could be fine-grained at the top but have a coarse, aligned crystal structure at the bottom, with a temperature difference of 40 degrees C between the two levels ranging from hard frozen to nearly melting. Lew pointed out that equations for the strength of steel don't hold up near its melting point, either. Richard's project to describe the development of first-year ice through the course of the seasons could tease out some of the details needed before a quantitative understanding would be possible.

For two golden years Richard lived in a Quonset hut at NARL with Arlene, Patuk, and baby Alice, gathering data all winter from the ice right outside and working in the labs and freezers next door. "What more could a guy ask for?" he said. Being on the ice all year gave him a great advantage over researchers dropping in now and then to take measurements. All through the dark of winter Richard watched the ice grow by the day, measured its temperature profile and thickness, and explored its terrain as his private domain. He saw its changes, not just discrete points along the course of change. In the spring he watched it melt and examined the brine channels. He took drill cores into the big freezers at NARL—mostly used by then for storing whale meat—and cut the ice into thin sections to mount on slides and view under polarized light. That process brought out the spectacular patterns and colors of large, interlocking crystals whose size and orientation he could measure. Richard needed the stable freezer temperature for his samples, which would change in outside air, but some people couldn't get over the irony of an Eskimo spending his days in a freezer examining ice. "I'd bring a few people in there and they'd say, 'What the hell are you doing?' " Guests could appreciate the beauty of the slides, but the fact that Richard was Iñupiaq didn't make his scientific interests any more intelligible to them.

One day Richard found a stretch of ice, a refrozen lead, that had been bent into a series of sine-curve-shaped hummocks with liquid puddles lying in between them—puddles that were liquid at an extremely cold temperature when they should have been solid. Richard took brine from the puddles home to measure its salinity, but it maxed out his instrument, even when he cut it with water and cut it again. Finally, he established that the brine was one-quarter salt. The next day, the cup of brine he had left on his windowsill was pure crystallized salt. As far as he knew, no one had observed this phenomenon before, but when he asked elders about it they weren't surprised: that was how they gathered salt in the old days.

Ultimately, Richard abandoned the degree with only the writing of the dissertation left to do. Partly, life became too busy with 'Berta's arrival and the gas field project to finish. But he also became uncomfortable with the idea of having the degree at all. He saw a lack of Iñupiaq humility in the basic assumption of his project, the idea that he could take traditional knowledge to a higher, scientific level. Through two years of study, he had discovered how little he really knew. What he had learned instead was that traditional knowledge existed as an organic part of a person living in the environment, a whole world constructed from experience, and couldn't be extracted and rationalized into data points. "I didn't want to become the ice man, the expert in a town full of experts, some kid from California that thinks he knows everything," he said. "To me, it's not so much about finishing a degree as continuing to learn about this life."

As he made that statement, Richard stood on white ice in pale sunlight, gazing over the sea from inside the hood of his white hunter's parka. We snacked on a frozen caribou haunch. The water was rough, so Savik Crew was not boating, instead just waiting for a whale to surface nearby, the harpoon and shoulder gun laid out with care at a high point on the ice edge. Waves boomed and reverberated underneath. Richard had moved the snowmachines back a little; the camp with the tent was well back among the multiyear ice. My questions were a distraction from his quiet watching until I asked one that really interested him, made him think: which way did he know more about ice, as a scientist or as an Iñupiaq? He debated with himself a bit before he answered: He knew more as an Eskimo. Scientists, he observed, know a collection of facts about ice; Eskimos know ice itself. "The best ice scientist is almost an Iñupiaq," he said. "If he's a good ice scientist, then he's thinking the way these people here do."

Hours later, on the way back to the NARL, I met Lew Shapiro for the first time, just ran across him a few hundred yards off the beach studying ice, and introduced myself. I looked at the sky with the feeling I was living in an improbable script—what was the chance Lew Shapiro would be in Barrow studying in this particular place on this particular day? But Lew didn't seem much surprised that I already knew all about him. Ice science is a small world; Barrow is a small town. Along with his colleague at the site, Dave Cole, of CRREL's main lab in Hanover, New Hampshire, Lew showed me amazing tablets of ice. Six one-inch-thick ice panels five feet tall and two feet wide stood in two perpendicular rows like ingenious works of modern art, the late yellow sun catching their flaws and channels in glorious colors. The scientists and their colleagues (including Richard, after he quit the Ph.D.) invented the machinery to create these panels. A seven-foot-long chainsaw mounted on metal tracks on an Eskimo-style sled swiveled down and made two long, precise parallel cuts. Working like glaziers, the men then lifted the ice windows from these slots and set them up vertically to examine and photograph. Data loggers at the same site kept track of ice thickness and temperature through the year, although in this abbreviated winter that had been possible only since January.

The first ice panels opened a whole vista of discovery and new questions. They showed, for example, that brine channels run diagonally before meeting and descending vertically; they form treelike patterns spanning meters, with the diagonal angle apparently determined by the temperature. At first, other experts thought the panels must simply be crooked, and still no one had explained why the brine should flow first at an angle and then vertically, but with one look it was indisputable: you could clearly see the elaborate plumbing of trunks and branches. The panels also showed the layers of ice—up to sixty horizontal layers of different colors and textures that formed over a two-hundred-day winter—and vertical alignment of ice crystals, too, apparently running in the direction of the current.

I felt a bit of vertigo thinking of the bottomless task of sorting out so many crystals arranged in patterns running through so many dimensions, including the dimension of time. But Dave Cole obviously relished the puzzle in its exquisite complexity. His goal of understanding the physical properties of the ice and how they developed through the year would require him to grasp ice behavior on a continuum of scales ranging from individual crystals to ice sheets, and from the perspectives of optics, mechanics, electromagnetics, and the chemistry of the organic gases in the pockets. As daunting as the project seemed, he had made good progress with this approach of starting from the tiny to understand the large. For example, he and his colleagues had learned that brine channels, in the odd way they gather in patterns, drill the sea ice with holes up to four inches in diameter that act like perforations on a piece of paper marked "tear here." Ice tends to break along those perforations; using hydrophones, the scientists had recorded the rhythm of cracks breaking in series from one vertical channel to the next. Finding out why and where brine channels formed could tell you something about how and where flat ice would break, something the Iñupiat did not know. It also would help calculate the strength of pieces of ice of intermediate size (the strength of larger pieces is governed by other factors).

The Cole-Shapiro team was an odd couple. Lew came across broad and blustery, talking with his body and a strong New York accent. Hollywood would never cast him as an Arctic scientist—they would give him a cigar to go with his bald head and big features and make him a political boss. Dave, on the other hand, looked the parsimonious New Englander he really was. His frame was tall and thin and his face long and narrow; his clothing was always in perfect order and his speech precise and thoughtful. These two had spent a lot of time together that winter—too much. First seven tough weeks camped out in Antarctica getting little sleep, with a storm that flattened the tents and damaged the equipment. And then here. Lew kept talking about quitting fieldwork, but he had been retired twelve years already and was still here—Matthew Sturm said Lew had quit more times than Michael Jordan.

Dave, like so many other Arctic researchers, missed his family: his wife, Karen Henry, also a scientist, and his two busy, athletic teenaged boys back in New Hampshire. He should not have been away from home so long this year, he said, but as a grant-funded scientist he tended to write extra proposals to make sure he had enough work, and this year he had won more grants than he expected. The family was angry. First they went out and bought a new computer, then they made an ultimatum: no more proposals for fieldwork without family consultation. Lots of polar scientists had lost their marriages by spending too much time away from home. By this stage in the trip, Karen and the boys had adapted to Dave's absence—he didn't exist back home anymore—and even his phone calls came as an intrusion in their impossibly busy routine. As Karen said, "I'm not very pleasant when he calls home. It's like, 'I really don't have time to talk to you right now.' " Besides, FedEx had lost Dave's $20,000 microscope, leaving him without a critical tool he needed to work effectively in Barrow (it turned up much later in the southern United States).

Through it all, however, Dave's professional cordiality traveled with him like a virtual seminar room, a collegial aura that pervaded his freezer laboratory and cafeteria table. As the master of capturing images of ice—beautiful images with bizarre colors and powerful, angular shapes—he was dedicated to spreading an awareness of what ice really looked like to the many scientists studying it from their own, narrow perspectives. He wanted modelers, oceanographers, biologists, remote-sensing specialists, engineers, and all the others to know the anatomy of a brine pocket, to see the intricate but regular branching of brine channels through the interwoven layers of porosity. Dave had long recognized that no one person could grasp, let alone solve, the entire sea ice puzzle. The problem crossed scales and processes. The physics and biology of brine pockets interacted with temperature and with time, affecting how the ice grew and broke, how it looked from a satellite and how it might respond to and influence climate. Ten years earlier Dave and his colleagues successfully calculated the strength of a thirty-meter-square section of ice in Antarctica that they cut loose and left floating while instruments stressed it and measured its responses. But to develop general equations with predictive power in the Arctic would require bringing together many more complexities and doing it in a system that was itself in a state of rapid change. Many specialists together would need to learn ice as a system rather than as their own individual baskets of facts.

In the NARL freezer Dave taught Matthew Sturm to cut thin sections from ice drill cores so he could make the images. When the passive microwave sensor passed over on the satellite in a year's time, Matthew wanted to make these pictures to put beside images of the same crystals recorded from space. Through the accretion of many views at many scales and analysis by many minds, a whole understanding of the ice could eventually emerge. The approach reminded me of the way the Iñupiat know ice, but using the instruments of science and the precise language of mathematics in place of the Iñupiaq language. That was the dream, at any rate—that a scientist's data could leave his or her filing cabinet, live beyond a career, and contribute to more than a collection of journal articles.

My original plan had been to compare Richard Glenn and Dave Cole. On the surface, it was tempting: two men of roughly the same age, working on the same problem in the same place with the same unusual colleague, Lew Shapiro. They even worked in the same freezer—a dreary chamber of dark metal that somehow felt colder inside than the equivalent temperature outside. Dave recalled how on his first trip to Barrow, when a lot of polar bears were hanging around NARL, the freezer reeked of whale meat, which transferred to his clothes; when he came outside he felt like he was walking bear bait. That seemed like a neat point of comparison, too: for Richard, going into the freezer was strange; for Dave, the challenge was coming out of it.

But the analogy didn't extend to the work itself. Richard's work as a graduate student a decade in the past bore little similarity to the deep and sophisticated study Dave pursued with advanced tools and the skills of an experienced experimenter in the prime of his career. And their role in the world was different. Dave would go home to a colonial farmhouse on four and a half acres under the broad-branching trees of a quaint New Hampshire village, where in the summer he could launch his boat on the Connecticut River just below the house and in the winter ski from behind the barn over twenty miles of groomed trails traversing the glorious wooded hills. Work would stay at the lab in Hanover, near the idyllic campus of Dartmouth College, the sea ice contained in freezers, a fascinating and rewarding but ultimately academic interest. The world would certainly benefit from the science, eventually, but the leap from a picture of a brine pocket to the coming global change was too great to provide a real sense of urgency. He was free to choose to spend less time in the field and more time with Karen and the boys without feeling he was letting down society as a whole.

Richard could have had a life like that, but as a young man he chose to take his place among the Iñupiat. His remarkable capabilities didn't belong to him alone anymore, his people also had a claim. Knowing ice and how it changed was more than a fascinating puzzle, it was a matter of life and death. To be a scientist gazing at ice crystals under polarized light, to gain that Ph.D., seemed an indulgence in the context of his world.

One day in Oliver Leavitt's whaling camp we talked about Richard in his absence. Oliver told me Richard would one day be the head of ASRC, among the Iñupiat's highest leaders. Oliver was irritated when I suggested Richard might instead prefer to pursue his interests as a scientist. "Most of them don't do a hell of a lot," he said. "He can do a hell of a lot more, if he wants to do more for his people.

"If you want science, you can buy it. But you can't buy leading your people."

Members of Savik Crew stood lookout. When the water calmed, they went boating. At night some lay down to sleep in the tent, the radio playing softly—rap music one night, another night some raving radical from Berkeley's Pacifica Radio. The support crew in town left hot meals at Niksiuraq, the end of the road, for a crew member to pick up by snowmachine. The best was aluuttigaaq, stir-fried caribou chunks in thick gravy. Days went by as we lingered at the ice edge, watching quietly, talking, watching; there were polar bear, seal, beluga whales. Roy went boating. Richard went boating.

One afternoon Richard saw a whale in the distance and quickly launched the boat with Eben and Benny, leaving me alone at the ice edge. Minutes later I saw a big bowhead right near the lookout, but the boat was long gone. I paced around. The sun was out but the temperature was around 10 degrees F, with a fresh wind. Even though I was covered head to toe in heavy Arctic gear that had kept me warm far below zero before, I had to move to produce enough heat.

A little old man appeared. He wore blue jeans and rubber boots that looked like they came from a thrift store, and a thin jacket and cap. My first thought was to try to protect him, that somehow he had wandered out here without proper clothing. He spoke English with a thick Iñupiaq accent and a lack of teeth that made him even harder to understand, but I gathered he wanted to know what was going on. I told him about the whale I had seen near the ice, and he nearly jumped with excitement. He said, "I'll get my boat," and ran off with tottering little steps.

When Savik Crew returned, they said I had met the great Malik. The follower and catcher of so many whales, the one who climbed on the whales' backs with an unerring lance, the one whose name was much too great to be applied to me. I was glad to have met him, glad he was a cheerful, friendly little guy, glad I had helped him, even if it hadn't been much.

Normally whales would be seen every few minutes, but now days went by without a sighting. The boaters went farther and farther afield—George Ahmaogak's crew once ran the risk of getting isolated on the far side of a huge floe. Those whaling the old-fashioned way to the south, like Oliver Leavitt, complained that the aluminum boats were scaring off the whales. But it could also be noisy snowmachines running back and forth or a too-wide lead spreading out the whales, or perhaps the main pod had simply passed by during the stormy weather. One afternoon a teenaged member of ABC Crew named Marcus Brower came to our lookout advancing the theory that the whales were spurning Barrow because the skull and ribs of the first whale—the only one caught so far—had not been pushed back into the sea for the spirit of the whale to return. With so many polar bears around the butchering site, the crew had chosen to retreat instead of cleaning up the leavings. "That's the only reason no one has gotten another whale," he said.

With the bad ice behind them, whalers frequently pulled camps back. After each pullback, more crews showed up at the north end to use the ramp chipped by George Ahmaogak's crew through the ridge at the ice edge. Oliver's resistance to opening the whole lead to the aluminum concentrated whalers here, north of Nuvuk. Normally only the crews who help build a trail and those crews closely related to them would share, but normally ice conditions did not make safe access to the water so difficult. Besides, George was by nature disinclined to say no, and I speculated that with his mayoral reelection coming up in the fall he was being especially friendly.

On May 7, the big ice pan near the ramp looked like a campground, surrounded by white, canvas-walled tents. Richard and Roy brought Savik Crew out, but they had limited time to hunt with Patuk's graduation coming up and visitors arriving from out of town. Roy went boating and Richard went visiting. In the warm evening light he joined Julius and Brenton Rexford at Atqaan Camp, climbed their nicely carved ice steps to a high lookout point, and talked about the conditions. A call came over the VHF radio that a current had kicked up and crews down south were pulling back. Here the water looked good, but soon it started flowing fast toward the north, as fast as a river. Richard called Roy in. A string of boats pulled out of the water at the ramp, captains quickly disarming their weapons, many hands reaching to help load boats on sleds. The tent that had just gone up came down again. The crew went back to town. No more whaling until after the graduation.

Each spring at a certain moment the sea ice begins to fall apart. The brine, stored in pockets and porous layers over the winter, hits a critical temperature and flushes out over the course of a few days in one great flow, drilling the large vertical channels that can act as perforations to crack the ice. Biological activity picks up, too, eating at ice like worms through wood. The Iñupiat were used to working on deteriorating ice; whales, light, and rotting ice all came together in the spring. But this spring the rot came faster and earlier than many could remember, and the ice was weak to begin with, with few anchors and too few cold months in which to harden. By the time George Ahmaogak's crew landed their big whale, as I described in chapter 1, the ice was wet and the temperatures breaking well above freezing every day.

The same day that the around-the-clock effort of butchering that whale was completed, more were caught. David Leavitt's crew struck a whale and killed it with help from other crews, and seven or eight boats joined to pull it to the ice, each attaching bow and stern to a long, heavy line tied to the whale's tail. J. R. Leavitt's crew got their whale to the ice and tied up at the Ahmaogak ramp, but with so much happening elsewhere, and so many workers resting after the Ahmaogak whale, they were short of workers to pull it out of the water. This was a new crew handling their first whale on the ice. Patuk, done graduating, went out with four other members of Savik Crew that evening to help. They were greeted with cheers by about fifty people already there. At the cries of "All hands!" and "Walk away!" everyone pulled, working hard. But the whale budged only a little and they couldn't hold it; it kept slipping back.

The whalers were on the seaward side of a long, wet crack, but everyone thought it would hold. Russ Page saw no cause for concern from his vantage at the National Weather Service ice desk in Anchorage: a low-pressure trough affecting the area could produce an ocean current, but it was too weak to warrant an advisory of a potential ice break-off. Richard Glenn, however, worried as he sat in Barrow listening to the discussion on VHF channel 72, thinking of Patuk and the other less-experienced Savik Crew members out on the wet, rotten ice in the warm night. He heard whalers at the southern section of the lead talking about rising water and a hastening current. Five years earlier he had decided to stay on the ice in questionable conditions despite Roy's advice and he had been wrong and floated out to sea. He heard on the VHF that the whalers Patuk was helping disagreed about what they should do—keep pulling or tie off the whale and retreat. His reading of personality guided him as well as if he could see the ice. "As soon as I heard squabbling was going on, I called them in," he said. He dialed his sister-in-law Ethel's cell phone and told the crew to come home. They didn't want to come, but he insisted.

"I felt really tempted to just stay," Patuk said. "It's just really hard to leave a crew out there when they didn't have that much help, and they were so happy when we first got there. They were celebrating that they got their first whale. We had to go up to the whaling captain's wife and tell her we were leaving because of the ice. And she tried to be understanding, but after we left she started crying. I felt really bad. I wasn't really sure the ice was breaking, but I had to trust my dad."

Patuk and the other Savik Crew members drove their snowmachines off down the trail, followed by Patuk's cousin, Justin Gatten, who had graduated with her. He stopped to talk to someone headed out toward the whale. In that few minutes' delay, he was cut off. As he traveled on down the trail the ice shattered around him. He kept driving, jumping small cracks until he reached one too wide to cross and parked his machine. With two other men he leaped from one ice chunk to the next, clearing water up to five feet wide before making it to the shorefast ice. Then he called his crew on the radio and gave them the news that they were adrift.

Brian Ahkiviana, the hardworking seventh grader from Oliver Leavitt Crew, was helping to pull up the whale, along with Jens Hopson, when the people around him suddenly panicked, yelling and running, and he ran with them. Snowmachines zoomed off chaotically. In town, senior whalers and elders converged on the Volunteer Search and Rescue. Oliver Leavitt, the president of the organization, took to the air on the VHF to calm the people fleeing and get them to line up their snowmachines and form groups on big ice floes where a helicopter could pick them up. The largest group, almost sixty, was at J. R. Leavitt's whale, and another thirty or so were grouped in an area about seven miles away. Oliver's son, Billy Jens Leavitt, along with Hubert Hopson and Gilford Mongoyak, were still out helping tow David Leavitt's whale, while their unattended camp and gear were floating away back on the ice.

The break-off followed the wet crack that people had hoped would hold, unzipping just outside the big pressure ridge that ran north and south like a levee, the boundary between the more solid, flat ice nearer shore and the outer rubble ice that had been glued together late in the season. The grounded pressure ridge had acted as a hinge point as rising water flexed the weak rubble ice, and now that ice was moving out from shore and disintegrating with ninety people on it.

The Volunteer Search and Rescue notified the borough's pilots, who went to their immaculate aircraft hangar to fire up the two big Bell 214 helicopters, each with two pilots and a crew chief. Oliver and the other volunteers coordinated over the VHF, directing the pilots to groups they had told to congregate. Navigating with a hundred-foot ceiling and fog and blowing snow that cut visibility and eliminated the horizon, the helicopters gingerly put down on the ice, keeping their rotors spinning to avoid applying their full weight. Fifteen passengers boarded each flight, with women and the youngest whalers going first. Brian and Jens were picked up within an hour of knowing they were adrift. By that time, they were already three miles out.

The boats pulling the David Leavitt whale, thirteen miles north of the point, had been towing for seven hours, and for the last four hours they hadn't moved; in fact, they were going backward. The current dragging on the whale was stronger than the outboard motors in the boats. When word came that the ice had broken off, the boats cast loose one by one and let the current take the whale away. Among those boats were Oliver Leavitt Crew and Hopson 1 Crew, whose camps were in the same general area off NAPA, where Jacob Adams's camp also stood—Jacob is the president and CEO of ASRC, as important a man as Oliver and also his close friend. The Oliver Leavitt and Hopson 1 boats drove toward camp, stopping by on the way to see the big crowd of people at the J. R. Leavitt whale. Order had been restored. People were standing together calmly on a big ice pan. Their fifty-two snowmachines and forty sleds were lined up, ready to be strung with a cable like beads on a string that the helicopter could pick up. Back at the three crews' own camps a chopper landed but the crews sent it away—they had their boats and intended to carry their own gear to safety.

The three crews pulled their aluminum boats out of the water, packed their camps, and took off together for the widening gap, about three quarters of a mile away, with a dozen men, a dozen snowmachines, and about ten sleds heavily loaded with gear and boats. They worked at a frenzied speed, but efficiently, and still with humor—Clayton Hopson said his snowmachine had to go first because it was brand-new. At the lead, they relaunched the boats—little Lund aluminum boats such as would commonly be used for lake fishing—holding the boats' sides firmly to the ice with ropes while driving a snowmachine onto each boat. They drove across just far enough to park the skis on the far gunwale and the track on the nearer one, finding the exact, precarious balance to keep a 500-pound machine perched high above the little boat's center of gravity. The three boats crossed a 300-yard lead of open water more than twenty times to get all the machines and sleds across.

On the far side, they still weren't on firm ice. Underneath them, rubble ice disintegrated as they drove, cracks opening ahead, behind, and on either side of the trail. A crack opened right under Perry Hopson and he fell off his machine and into the water. His cousin Clayton jumped off his machine and grabbed Perry with one hand and held his machine with the other, to keep it from sliding and sinking away on the broken ice. He was able to hold on just long enough for others to grab hold and pull up the man and the snowmachine. They didn't pause. Just to follow the trail they had to cross one newly opening crack after another. Finally a large crack blocked the way with a span of about two feet of water that was spreading fast. Each man unhooked the sled from his machine, circled for a running start, gunned the motor and jumped the crack. Then they jumped back, attached ropes between the sleds on one side and the machines on the other, and pulled the sleds across the crack one by one, their long wooden runners bridging both sides. As the crack widened, the process was trickier with each sled. To the last sled they attached two snowmachines. The front of the runners made it over, but the back of the sled dropped down into the crack, trying to pull the machines with it into the water. With the full power of both machines fighting the sled's weight, it rose and came across.

The three crews parked in safety on the far side of the big, grounded pressure ridge and watched the ice they had just crossed break into ever smaller pieces, shearing off even closer to the ridge. They watched as the helicopters passed overhead with half a dozen snowmachines at a time hanging under them and one with a skin boat, broken in half. From the time they released the whale to standing on safe ice only two hours had passed.

The work of saving whalers and gear went on all night and into the morning. Oliver stayed at the Volunteer Search and Rescue coordinating the flights. Many other whalers were at his side helping in any way they could, including Richard Glenn. A helicopter saw Justin Gatten's snowmachine and a machine belonging to one of the men who had been with him drifting off on a small piece of ice. The ice flipped and the machines disappeared into the ocean. Late in the night a drunk went on the VHF and started ranting and yelling, disrupting communications. Oliver said that happened often, whenever someone drank too much and wanted to spew grievances to the world, but authorities had been unwilling to take action against the offenders. Tonight, with all that was on the line, Richard grabbed a hacksaw and climbed up the outside of the drunk's house to the roof and cut down the antenna.

By midmorning, everyone was safe and most of the whalers' gear recovered, thirty sling-loads of it, including as many snowmachines and sleds as people. When the last load left the ice, it had rounded Point Barrow and was headed east at three or four miles per hour. Both the whales were lost. Richard had the satisfaction of knowing he had pulled in his crew before the break-off. He shrugged off my question when I asked if this made up for the time in 1997 when Roy wanted to pull back and he put his foot down. Family relationships rarely change so easily.

Billy Jens Leavitt headed back out as soon as conditions permitted. Barrow caught one more bowhead that season, a huge whale. Helpers crossing the ice to butcher it had to drive their snowmachines through deep water. Soon the last whalers gave up hunting. Normally whaling goes on into June and the ice doesn't go out until July, but this spring the ice became too wet and dangerous in May. The total catch of only three whales was far too little to sate the community's appetite; some years, they brought in twenty or more.

At least Barrow still had fall whaling to look forward to. The geography of Point Barrow brought the bowhead migration near shore in both directions. In fall, when there was no shore ice, Barrow crews pursued whales in twenty-foot boats with large outboards. For most other communities along the Arctic coast, however, fall whaling wasn't possible, so the unsuccessful spring whaling season meant little or no maktak for an entire year. Point Hope, one of the most traditional whaling villages—the women there still brought freshwater to the dead whale—caught no whales at all that spring for its 760 residents.

The village of Little Diomede, about 150 people on a small island in the middle of the Bering Strait, wasn't able to whale for bowhead that spring, so on June 25, a group of seven aluminum boats went out in open water for a gray whale, a more dangerous species with barnacles that spoil the maktak. A bowhead, like any other animal, can turn on a hunter if it feels cornered and can even crunch an umiaq in its long mouth, but that is very rare. Bowhead are usually placid. Gray whales, on the other hand, are smaller, more maneuverable, and known for fiercely ramming whaleboats or striking at them with their tails. Their Alaska population was stable and numerous enough to support a hunt, but the Iñupiat rarely pursued them and did not have a legal quota. The Little Diomede whalers were desperate enough to try anyway.

Gray whales were ample that night. After midnight the hunters chose a small one and hit it with a harpoon. Several hours later, a second boat moved in to throw a second harpoon at the same whale, the Ozenna family's boat. But the whale struck first, rearing up right underneath the boat and flipping it with its tail. The other boats quickly retrieved all four men from the water, but one, Melton Ozenna, was unconscious and bleeding from the ears. On the way back to the village crew members got the water out of him and started his breathing, but before they reached shore he fell unconscious again and never woke.

The International Whaling Commission imposed a fine on the Little Diomede whalers for whaling without a quota, but Maggie Ahmaogak said the Alaska Eskimo Whaling Commission would pay it. She said, "They're desperate to feed their people. They're desperate to have meat for their village."

CHAPTER SIX

The Supercomputer

A TUBE atop a fifty-five-foot tower over a lone building north of Barrow sniffed the wind arriving from the Arctic Ocean. The way there led beyond NARL, down a long, gated driveway and past the air force radar station. In failing light under a low overcast, the table-flat plain of snow faded into the sky as a single, colorless blue-gray, without features or depth, upon which this squat rectangular box of a building seemed to float out of context. As a work of art, it would be a sterile abstraction. Unable to find a pathway in the flat light from the road to the front of the building, I climbed through a snowbank to the steps and knocked on the door. Humanity was in the process of deciding the future of all living things through its use of energy stored in hydrocarbons. How fitting it seemed that this outpost, where the information critical to that choice was gathered, balanced in a shapeless void on the edge of the world.

Dan Endres opened the door with the diffident smile of someone who doesn't get a lot of visitors. The guest book confirmed this impression: a few pages covered many years, with only a name every few months from September to May. Inside, the National Oceanic and Atmospheric Administration (NOAA) Climate Monitoring and Diagnostics Laboratory (CMDL) was an ordinary office and scientific workshop, brightly lit by fluorescent fixtures, the walls and ceiling covered with sound-absorbing foam, racks of machines clicking, humming, and thumping around the edges of one big room. Dan had been sitting at a desk in a hallway near the door facing a screen of e-mail. He was used to being here and used to being alone. Growing up in Colorado, he chose between physics and the saxophone, and physics looked easier. A bachelor's degree in physics qualified him to paint houses, which he did until 1981 when he got a job calibrating climate lab equipment in Boulder for three years. "This position came open, so I came up here. It was going to be two or three years to put some money in the bank for school. But that never happened. Seventeen years slipped by and I'm still here."

Dan was a cheerful, laid-back guy with a slight paunch covered by a fleece pullover, but he seemed defensive about what I might think of a single white man living so long in Barrow tending machinery. He hastened to portray his Barrow as a hobbyist's paradise, the winter's sixty-seven-day night as an opportunity to play cards and dine with friends, to shape impossibly delicate wooden cups on a lathe, which he sold through a gallery in Colorado, to assemble stained glass windows, carve knives out of steel, and spin ceramics. In a class at Ilisagvik College he helped weld together a twenty-foot aluminum boat. He had become a well-liked scientific elder in town, with tenure long enough to know what had been done before and to help guide BASC on one of its committees. All true, and his sacrifice was laudable, since management by a single operator helped ensure the quality of the long-term data sets—people miles above Dan in the NOAA organizational chart valued him for that—but still it seemed an extreme sacrifice to spend an entire career at such an outpost gathering measurements to be used by others.

The first machine Dan showed me in the room of clicking and humming machines was a metal box covered with gauges for measuring freons, halons, and other ozone-depleting chemicals. As he started to move on to the next device I stopped him and asked how this one worked. He paused. "You mean, how it really works?" I nodded. A smile broke across Dan's face as he opened a panel and began an explanation of how beta decay from nickel 63 would lend tracer electrons to individual gas molecules, which could then be counted in an electric field. Since a 1987 Montreal international protocol banned chlorofluorocarbons such as freon from air conditioners and the like, Dan had watched as the levels on this machine trended down nicely. Here was the joy of a tinkerer, the pleasure of understanding a complex piece of equipment and making it work to learn something real about the world. He knew these machines well enough to read in the carbon data the signature of a snowmachine or of a Search and Rescue helicopter passing miles upwind of the lab. I realized I was wrong about his sacrifice. Dan Endres was in exactly the right place.

The lab was the best equipped in the Arctic for measuring climate change indicators. It was the tip of a long air-sniffing antenna reaching from the globe's inhabited zones to this perimeter of the human habitat. Numbers flowed to scores of PIs with individual projects all over the world and to thousands of researchers trying to grasp the climate as a system, mathematically, and predict its future. Dan was the keeper of the longest continuous record of atmospheric carbon dioxide in the Arctic and the second longest on earth after the record that was started in 1958 at Mauna Loa, Hawaii, by Charles Keeling, discoverer of the rapid increase caused by human activities (a project funded in part by the International Geophysical Year). Dan kept handy a plot of his lab's entire twenty-eight-year record of atmospheric carbon dioxide to show visitors. He said, "That's the most famous data to come out of the Arctic anywhere at any time." It was an extraordinary graph, almost too clear and unequivocal to be true. Each year, it showed atmospheric carbon dioxide rising and falling with the season, but with each winter peak higher, in an inexorable stair step up the page. Winter temperatures trended upward along the same path, one degree C per decade.

The relationship between temperature and atmospheric carbon dioxide has been understood since the nineteenth century and is conceptually simple. The sun's energy, which powers the weather and virtually all life, arrives at our planet in short wavelengths, such as visible light, that pass through the atmosphere relatively easily. But when that energy reflects off the earth's surface it transforms into radiated heat, or long-wavelength infrared energy. Certain gases in the atmosphere that are transparent to short wavelengths—including water vapor, carbon dioxide, and methane—instead absorb long wavelengths. The sun's energy bouncing up from the earth heats those gases instead of escaping to the stratosphere or out to space. This warms the earth and the lower atmosphere, called the troposphere, and cools the stratosphere. The phenomenon is called the greenhouse effect because glass in a greenhouse works essentially the same way: it lets energy come in as light but won't let it go out as heat. The greenhouse effect is powerful, the second most important factor after the sun in determining the earth's temperature; without greenhouse gases, the planet would average –18 degrees C, too cold for most life (the current average is around 14 degrees C). Carbon dioxide heats up so readily under infrared energy that instruments built to detect minute quantities of it, including the one at Dan Endres's lab, do so by exposing air samples to infrared and measuring the change in pressure.

Besides being a greenhouse gas, carbon dioxide is necessary for life. Plants capture energy from the sun by combining the carbon atom in CO_2 with water's one oxygen and two hydrogen atoms (H_2O) to produce sugar (CH_2O) and an extra O_2 oxygen molecule that returns to the atmosphere. That's the process of photosynthesis in its simplest form. Animals make use of the binding energy in the sugar molecules by reversing photosynthesis, combining CH_2O with O_2 from the atmosphere and emitting CO_2 and H_2O, the process of respiration by which we live, which is also the basic chemistry of fire and decomposition. The earth's plants and animals constantly exchange the same carbon, oxygen, and hydrogen atoms back and forth; we spend only the binding energy of the sugar molecules originally captured by photosynthesis. That's called the carbon cycle.

Air bubbles trapped under Antarctic ice showed that for the last 420,000 years the carbon cycle was in a rough range of balance, with carbon, oxygen, and hydrogen exchanged between plants and animals in generally equal quantities. The carbon in the atmosphere seesawed regularly from 180 to 280 parts per million. As it did, temperatures rose and fell through glacial periods and intervening warm periods in fairly close correlation. In the last fifty years, however, our human activities raised the carbon dioxide in the atmosphere well above any they attained during those hundreds of millennia.

We had good reason for doing that. We outgrew the amount of energy captured by the plants living around us. Much more energy was available from ancient photosynthesis, in the form of coal and petroleum. It allowed societies to stop killing whales for light, stop deforesting land for heat, and stop using human beings and animals as machines for work. Fossil fuels saved the environment. And they made life immeasurably easier and more fulfilling for people freed from the limits of the energy available in their immediate surroundings. Don't try to tell an Eskimo elder that life was better before fuel oil heaters. Burning blubber and seal oil didn't work as well; the Iñupiat suffered in cold in frame houses, and in their more energy-efficient sod houses they developed chronic lung disease from childhood because of the smoke they breathed. Besides, it's doubtful today's population on the North Slope could be sustained that way, just as the balance of the world's population in cold and temperate regions has grown too large to heat with wood. Transportation using fossil fuels also improved life vastly. Not many people remain alive in Western society who remember making the switch from using animals for transportation, but plenty of elders in Barrow were around in the 1960s when snowmachines took over from dog teams. Suddenly, instead of taking a day to get to hunting camp, you could do it in an hour or two and use that day for something else. No longer did you have to fish and hunt to feed dogs, or care for them all year round. The gifts of leisure time and greater freedom, the time to learn, think, and create, to accomplish more and go more places, all came in large part when we were released from the limits of the energy available from contemporary photosynthesis. The fact that many people wasted these gifts doesn't diminish their intrinsic value, as people who lived the other way can testify.

Once we started burning fossil fuels, however, the carbon cycle was bound to become imbalanced. Current levels of human energy use bumped up the respiration side of the equation by 8 percent. Fossil fuels multiplied the energy one person could use by many thousands. A human's daily diet of 2,000 calories equaled 8,000 BTUs of energy, enough to continuously operate a 100-watt lightbulb. A gallon of gasoline contains 125,000 BTUs of energy, more than the human body uses in two weeks. Burning one six-pound gallon of gasoline puts twenty pounds of carbon dioxide into the air, of which five pounds is carbon and the rest is oxygen (we'll just talk about the carbon, for simplicity). Wood is half carbon, so to recycle the carbon in one gallon of gasoline a tree has to grow by ten pounds. To recycle the carbon released to fly coast to coast round-trip on a passenger jet takes about 1,750 pounds of wood growth for each passenger. The total energy use of the average American loads five tons of carbon into the atmosphere annually. A growing forest (not a mature forest) can use about a ton of carbon per acre per year, so to balance the energy use of each American would require about five acres of young, healthy forest. That's far more forest than we have. More fundamentally, however, forests don't only grow, they also burn, age, and decay, releasing carbon back to the atmosphere; only if all the forests were cut down regularly and buried deep underground would they really negate carbon taken from underground and put into the atmosphere.

In 1850, atmospheric CO_2 was 288 parts per million; by 1958 it was 315 parts per million; as I wrote this, it was over 370 and rising at more than 3 parts per million per year. As a rule of thumb, an increase of 3 ppm in the atmosphere resulted each year from the 6 billion metric tons of fossil fuels burned on earth. But rules of thumb can be misleading; the true picture was much more complicated. The biosphere was absorbing more carbon than it did before fossil fuel use. Images from space showed a greener world; more carbon dioxide helped plants grow. Forests that were cut down for fuel and other uses were growing back: from 1950 to 1992, the amount of carbon stored in the forests of the eastern United States rose by 80 percent as formerly logged and farmed lands regrew. But when those forests reached maturity, they wouldn't soak up as much carbon anymore. Moreover, more forest fires, caused in part by warmer weather, were cycling carbon back out of the forests at a faster rate. The oceans were the most important cushion for our carbon emissions. Currently, ocean water dissolved about 2 billion metric tons of carbon a year from human activities. Phytoplankton floating on the surface used immense quantities of carbon for photosynthesis, too, some of which sank down deep into the ocean and out of the contemporary carbon cycle. But ocean chemistry limited the amount of carbon the water could dissolve. And the oceans' ability to dissolve CO_2 and to use it biologically declined as the water warmed. At a certain point, that drop could be severe.

Forests in temperate and tropical regions were like checking accounts for carbon: photosynthesis made deposits but fires and decay made withdrawals. Because these forests grew fast and covered large areas, the carbon on deposit at any one time was large, but over the long term it cycled through with debits and credits in rough equality. Deep ocean sediments that formed carbonate rocks such as limestone were more like permanent investments: once carbon entered, it was out of circulation until geological forces uplifted and eroded the rock. Arctic tundra and northern forests, called boreal forests, were somewhere in between—they were like savings accounts. Plants and trees there grew slowly and didn't capture much carbon in a year, but much of that organic material fell to ground that was cold and damp on top and frozen a little deeper, so the material would not decay, building up a positive bank balance over time. As long as it stayed cold, that carbon was mostly out of circulation. Over the eight millennia since the last glacial period this account had grown large. The first meter of soil under the Arctic and the boreal forest were thought to contain 450 billion metric tons of carbon, more than humans have ever released and comparable to about two-thirds of the carbon in the atmosphere.

It would be nice to know if the warming of the tundra would release all that carbon, but of course that wasn't simple, either. The rising trend of carbon dioxide on the graph Dan Endres showed to visitors to CMDL wiggled drastically up and down in time with the seasons. Similar measurements near the equator, which has no seasons, or in Antarctica, which has almost no life, rose on a smooth ramp from year to year without these wiggles. In winter, Barrow was dark and cold; there was no photosynthesis, but respiration went on, so atmospheric carbon dioxide built up, peaking in May. In the summer, plants used the carbon in the air and the CO_2 concentration dropped, falling as far below the tropical and Antarctic stations in August as it was higher in the spring. By October, all the stations were briefly the same again, inexorably 3 parts per million higher than the previous October. In 2001, the range of Barrow's wiggles was 16 parts per million from low to high. And the range of the wiggles seemed to be growing each year, too, whatever that meant.

Walt Oechel, an ecologist at San Diego State University, began studying how carbon enters and leaves the Arctic tundra bank in the early 1970s, around the same time NOAA's CMDL was beginning its measurements of atmospheric carbon. He was a member of a huge young group of ecologists who came to NARL around that time for the International Biological Program, an ecology version of the IGY. Thirty years later, some of those scientists were leaders in their institutions and fields, teachers who brought up generations behind them, training graduate students who studied changes on some of the same plots they planted when they were graduate students. The students and postdocs still tramped through the NARL Hotel, BASC, and Dan Endres's back door during long summer days of squatting over tundra hummocks among the mosquitoes. Walt had done especially well in his career, directing his own Global Change Research Group at San Diego, in part because he chose the right topic all those years ago—he set about trying to measure the transfer of carbon between the tundra and the atmosphere.

Walt started with small plots and simple experiments, but as the story developed he reached beyond the plant ecologists' typical horizon of a few meters square and set up equipment to measure the carbon dioxide from the whole landscape. He erected towers with equipment able to sense the carbon in vertical winds—eddies of turbulence spinning up and down that move across the landscape—recording how much was moving from the landscape to the sky or back down. He even equipped an aircraft with this equipment, flying hundreds of miles thirty feet off the ground while equipment sniffed for carbon rising or falling.

During the cold decade of the 1970s, measurements showed the pattern that researchers expected as tundra banked carbon in its cold, damp soils. But as the weather warmed quickly in the 1980s and the tundra warmed and dried, the direction of flow reversed and the tundra began emitting carbon—a lot of it. Extrapolating the phenomenon across the Arctic, the release amounted to 300 million tons of carbon a year, a full 5 percent addition to all the carbon coming from the world's burning of fossil fuels. Walt published his findings and made a big splash. But as the warming progressed in the mid-1990s, the carbon emissions slowed and at times even reversed, sucking carbon down into the tundra again. Today, the tundra is still a net carbon source over the course of a year, but not as much as it was two decades ago.

By capturing carbon within sealed chambers on the tundra with controlled temperature and moisture, Walt and his colleagues saw how warming and especially drying gave soil microbes a chance to burn long-dormant carbon—that was the increase in emissions in the 1980s. But the microbes also changed the soil chemistry to the benefit of plants, and in the 1990s the plants finally caught up, increasing photosynthesis and taking up more carbon. At the same time new, larger species asserted dominance, especially shrubs, which also increased carbon uptake. (From another perspective, that was the same change seen by Matthew Sturm and Thomas Itta, Sr.) In total, the microbes' respiration of the old carbon still won, but it wasn't as bad as it had been after the first wave of warming. That didn't mean, however, that the system would adapt again to further warming, taking up more carbon when more carbon is released. Maybe the next step up in temperature won't come with a mitigating increase in photosynthesis. Oechel's work helped prove that we don't know enough to make a prediction.

What we really wanted to know was what would happen next. That was the magic, the power and the reassurance of science, to make mathematical predictions about the real world—to calculate where a thrown ball would land—and to have those predictions prove true anywhere, at any time. That was the thrill of having an experiment work in an introductory lab course in school. It was also the fascination of a novel I read as a teenager, Isaac Asimov's 1950s *Foundation Trilogy*, in which Dr. Hari Seldon unlocks the laws of history and attains the ability to calculate the future mathematically. But as an adult, knowing more about the world, I find it difficult even to fantasize such a universe, and I know that no one will ever make Dr. Seldon's calculations.

At the scales at which we live (as opposed to subatomic scales), physical laws determine the outcome of events, and true randomness is hard to find. Determinism promises that two perfectly identical situations always turn out the same way, a starting point for predicting the future. But a scientist studying weather in the 1960s put an asterisk next to determinism that banished it from practical application in much of the real world. Ed Lorenz, of MIT, found that some simple systems of mathematical equations representing real physical laws produced drastically different results when initial conditions were nearly but not quite the same. Determinism was real, but in these systems it could predict the future only when the present was known perfectly in every detail, which was impossible. That asterisk next to determinism Lorenz called chaos.

If the idea seems abstract, think of dropping pebbles down a mountainside. No two end up at the same place, no matter how carefully you repeat the experiment. Indeed, based on a tiny difference in the starting point, one pebble could start a landslide and another could bounce into your shoe. The evolving shape of a puff of smoke or whether it will rain two weeks from now are similarly unpredictable, and not only because we don't know enough. We can't know enough. The mathematics behind interacting physical laws explodes tiny differences into big ones too quickly.

Even without understanding nonlinear mathematics it is possible to get a feel for how this works. On the normal scale of weather predictions, it takes two days for small differences in wind or temperature to double—two similar storms will be twice as different two days later. With possible futures diverging that quickly, weather forecasters with good information can reliably predict wind and temperature three days ahead. For faster-changing events on smaller scales, such as thunderstorms, the error doubling is faster. For persistent systems covering large parts of the globe, it is slower. Weather services extend the range of their reliable predictions by studying persistent systems, ignoring details, stating averages rather than particulars, and using maps of larger areas, but forecasts they produce ten days out are often wrong and so vague as to be of limited use. Forecasts two weeks out with any real specificity may be a physical impossibility.

On the other hand, many things are predictable practically forever. We know that spring will follow winter. We know that Barrow will be colder than Honolulu. We can predict broadly about some complex systems with certainty: Alaska's southern coastal mountains will have milder, moister seasons than the Interior, supporting a richer and more diverse biological community. This is the difference between weather and climate. Choosing shorts or long underwear on a particular day is about weather; the ratio of shorts to long underwear in the drawer is about climate. Weather happens in a particular place and time, climate happens in a place through a smudge of time, or a time through a smudge of space, and usually both.

We can't experience climate directly. Every moment of life happens in a particular time and place. The climate we perceive is a metaphor for the sum of weather conditions over a chosen span of time and space. Knowing an area's climate is like knowing a person: the sense of a personality develops through extended observation until it seems solid and tangible, but that sense is never more than an estimation and often has to be adjusted to account for new events. Ecosystems describe climate zones by where they grow. Weather statistics define climate using mathematical averages at particular spots. Subsistence hunters know climate through their experience of their home country. A climate exists only in the context of sufficiently broad boundaries, and boundaries must be chosen one way or another.

By setting the boundaries as broadly as is conceptually possible—the entire planet over hypothetical time—Syukuro Manabe and his colleagues at NOAA's Geophysical Fluid Dynamics Laboratory (GFDL) in 1963 built a working climate out of pure mathematics. Although the model lacked the dimensions of the real world or many of the processes that make the real world work, computer calculations did produce an atmosphere that looked right in important respects and that even answered questions about why the real atmosphere's layers occurred where they did. Three years later, Suki (as he is known) introduced a more realistic version of water vapor into the model—it is a powerful greenhouse gas—and produced the first meaningful prediction of how increasing carbon dioxide would affect temperature. A doubling of CO_2 from preindustrial levels would raise global surface temperatures 2.4 degrees C above the current 14 degrees. That paper was probably the most important ever published in the field. The work of the decades that followed progressively buttressed its basic conclusions. No one can doubt any longer that more carbon dioxide tends to warm the atmosphere.

Suki earned a reputation as the most delightful of men as well as a pathbreaking atmospheric scientist. We met in a little office he was using in retirement on Princeton's Forrestal Campus, the home of GFDL and a center for great brains three miles from the main campus, where the quiet was so thick I felt like tiptoeing even outdoors. Suki dispelled the hush. He still brimmed with excitement about his work and offered unlimited time to talk about it. He beamed, he jumped from his chair and bent his knees as if to lift up the big ideas, he passionately expounded on the next important discovery. Still speaking in a thick Japanese accent after more than fifty years in the United States, he joked and punned in stylish English, spinning out ideas with technical precision and rhetorical hyperbole.

In 1975, Suki published a model in three dimensions that included some of the details the earlier models had left out, including land and sea, the water cycle, snow, ice, the transfer of heat between the equator and the poles, and nonlinear mathematics. Like the earlier model, the three-dimensional model greatly simplified reality and looked only at climate equilibrium, not rates of change. But it was an important first, and it said that CO_2 doubling would raise the global average temperature 3 degrees C rather than 2.4, with the increase in the prediction largely caused by changes in snow albedo. Precipitation would increase and warming would be stronger and faster in the Arctic than elsewhere. In the decades since, those things did start to happen. That *was* something like Dr. Hari Seldon unlocking the laws of history. It remained to be seen, however, what would happen when climate modelers tried to go beyond this broadest, simplified scale to question nature more closely.

When I struggled through textbooks about climate modeling, I sometimes felt as if I had entered Asimov's novel. The audacity of the enterprise was similar. Climate modelers started at the beginning, with conservation laws of momentum, mass, and energy, the first of which could be found in Sir Isaac Newton's *Principia Mathematica*, published in 1647, as the second law of motion. A model had to break the atmosphere and oceans into vertical and horizontal grid squares, which, owing to the curvature of the earth, were geometrically complex—curved, narrower farther from the equator, squeezing together to points at the poles. So, to start, Newton's second law had to be expressed in terms of these curved spaces, with equations included to account for gravitation, friction, and the motion of the earth, all worked out in the three dimensions in space. Once the model was running, the equations in each grid square would take output from adjoining grid squares, the computer would perform the calculations and pass the resulting output on to the next square, simulating, if all was well, the movement of air or water from one place to another over a given step in time. But that was just the baby stuff. A full working model needed equations that reflected everything that affects the climate: radiative heating and cooling; convection; gas chemistry; energy absorption and transparency; atmospheric pressure; wind; turbulence near the ground; humidity; condensation; evaporation; freezing; snow albedo; sea ice albedo, brine exclusion, roughness, and movement; ocean currents and salinity; land topography and ocean bathymetry; freshwater runoff; soil moisture; land cover; and many other phenomena. And it needed equations that could transfer energy, momentum, and water between the atmosphere, ocean, sea ice, and land. It would be nice to include biology and chemistry too—with CO_2 feedbacks from forest fires, melting permafrost, oceans, and so on—but that was mostly a dream for the future. The "to do" list remains lengthy.

Modeling the climate occupied thousands of scientific careers and dozens of supercomputers all over the world and the challenge of the work led virtually indefinitely into the future. Field researchers, when you asked the ultimate purpose of their studies, almost invariably mentioned further refining GCMs (General Circulation Models—or Global Climate Models, which is easier to remember). The model had become, in its ideal conception, a summation of all knowledge about the environment. To some extent, the movement to relate everything to modeling was driven by funding—if you wanted to study snow, ice, or tundra, you needed to make the work relevant to climate change, and somehow contribute to modeling the climate, or your chances of getting a research grant were reduced. But the models also had become an organizing principle, their need for more precise equations for physical processes shaping both the choices of research topics and the way workers thought of their research. Rather than being dedicated to understanding the natural world in itself, a career spent studying a process in nature attained value for its potential to add lines of computer code to a GCM. Field studies often focused more on the needs of modelers than on the interests of individual scientists who were the true experts on their own part of nature.

Thirty-five years after Suki Manabe published his first climate change model, computer modelers had become a priesthood for the oracle of the supercomputer. Assessments of real world climate change impacts, published by the Intergovernmental Panel on Climate Change (the IPCC) and other bodies, rested on model predictions as their starting point. Consequently, the policy debate about climate change consistently came back to discussion of models, with some emphasizing their predictions a century into the future, and others their deficiencies and mutual disagreement. Although the basic physics of the greenhouse gas problem were easy to understand and difficult to dismiss once one grasped the scale of the forces involved, scientists willingly followed the path of increasing complexity as they tried to comprehend the entire climate system mathematically, marching far beyond the fringe of their audience's comprehension. For policy makers, the debate was reduced to competition between these scientific priests. Statements by organizations such as the IPCC emerged from the consensus of experts; climate change believers always won these votes. But policy makers were free to listen to whichever priest they chose, and there were always a few scientists somewhere raising objections over the models—an easy thing to do. These models were being asked to accomplish something no human creation had ever done: to predict, with some precision, based on first principles, what life would be like in a hundred years. The opening scene in Asimov's 1951 novel could have been written about a climate scientist predicting global change to a congressional committee, rather than about a mathematician of "psychohistory" predicting the disintegration of interstellar society to a prosecutorial tribunal:

Q: You are sure that your statement represents scientific truth?
A: I am.
Q: On what basis?
A: On the basis of the mathematics of psychohistory.
Q: Can you prove that this mathematics is valid?
A: Only to another mathematician.
Q: (with a smile) Your claim then, is that your truth is of so esoteric a nature that it is beyond the understanding of a plain man. It seems to me that truth should be clearer than that, less mysterious, more open to the mind.

A reasonable point, even in the mouth of an evil interlocutor representing a decadent galactic empire. Maybe the scientists didn't know what they were talking about. Indeed, their role as experts passing down received knowledge had allowed them to avoid embarrassing discussions with laymen on such matters as "flux adjustments" and "parameterization uncertainties." Flux adjustments were simply fudge factors; they were in decreasing use but still made some GCMs go. Essentially, modelers whose simulations went off track put a thumb on the scale in calculations showing the transfer, or flux, of heat and water between the ocean and atmosphere, adjusting the numbers until the model output became more realistic. Of course, once you started fiddling with the calculations it also became more difficult to claim you were modeling the real climate based on first principles. But sometimes a flux adjustment was the best way for a modeler to understand what a model was doing and to glean meaningful information from it.

Parameterization was a more fundamental issue, indicative not so much of fudging as of an inherent uncertainty about how the climate works. Since GCMs could not track every molecule being warmed by every photon from the sun, approximations of processes at larger scales had to do the job. Big approximations, because model grid squares were a hundred kilometers or more on a side (to make them smaller would require invention of faster computers). For example, the movement of sea ice depends on sea level differences and on winds and currents acting against the ice's strength and resistance to deformation—its mechanical properties. Dave Cole and his colleagues were still trying to figure out the mechanical properties of ice; as we have seen, that work was not simple, and it was far from complete. Parameterizations dealt with these unknowns either by leaving out the entire issue (which is not valid with something as important as sea ice), by specifying figures from a table based on observations (which told you nothing about how the ice might behave in a new environment), or by using simplified equations that the modeler hoped would be good enough. Some GCMs' treatment of sea ice moved from the second to the third of these techniques with equations that made broad assumptions that were not really true, including that ice was essentially moldable rather than elastic and that there was only one kind of ice, with one thickness. It was little surprise, therefore, that most models didn't work when it came to sea ice, or even for the Arctic as a whole. Some showed ice disappearing entirely in the summer, others showed it staying everywhere in summer; some had the seasons right but didn't show ice in the right places to match up with today's climate. To get a simulation to work, modelers could fiddle with parameterizations until the program behaved. No one could really complain they were cheating, because the range of potentially valid choices was so broad.

Nailing down the truth as it existed in nature to improve parameterizations sent scientists after unexplored details of how the world works. The prime example was clouds. When modelers included them they found that the physics of how clouds worked wasn't understood very well. Clouds were important. Water vapor is a powerful greenhouse gas, but when it condenses into liquid water droplets in a cloud it can shadow and cool the earth instead. A warmer climate would tend to create more clouds and that would decrease warming, a negative feedback. Fog is a negative feedback, too: when the ground or sea warms faster than the air, fog forms and cools the surface. But clouds can also have the opposite effect. Thick cloud cover can blanket the earth and keep it warm in winter: a cloudy Alaska winter night is usually warmer than a clear one. Modelers grappled with both the inherent complexity of clouds and a lack of knowledge about their details. Cloud physics had never been such a hot topic before.

Arctic climate modelers found GCMs had trouble showing the buildup of clouds in the central Arctic and thought the problem might lie in their use of a cloud model developed for temperate regions. The mathematics assumed that when humidity got to 100 percent, water droplets would condense into clouds; colliding droplets would become raindrops; the presence of ice crystals would simply speed up the process. That wasn't working. To get it right, atmospheric physicists set about trying to keep track of clouds' ice crystals and water droplets separately and at each gradation of size—a bunch of categories all together. For each category, they calculated many variables, such as the heat gained or lost through freezing and melting, the shape of the crystals, likelihood of collisions, likelihood particles would stick after collisions, and so forth. At each time step, the model would recalculate the fate of the droplets and crystals at all the sizes with all the associated variables. Adding so many details was costly in terms of time, money, and certainty. If the model didn't work, it could be the fault of any one of the new variables, or an interacting combination of them. Fixing one problem could create another problem somewhere else that might be even more complex.

Scientists looked for the arcane facts the models needed about cloud droplets and crystals in lab experiments, from airplanes, and at a metal-walled observatory that sat on the tundra next door to CMDL in Barrow, looking like something out of 1930s science fiction. This was where Bernie Zak contributed to the good of the world. As an atmospheric physicist with Sandia National Laboratories in Albuquerque, Bernie managed the Department of Energy's Atmospheric Radiation Measurement (ARM) site in Barrow (other sites were in Oklahoma and the South Pacific), measuring how energy passed through each kind of sky that passed over Barrow. Strange instruments on platforms shot the sky with radar and lasers and scanned at infrared, microwave, and other frequencies. A computer screen inside the narrow little building showed passing pockets of air turbulence a thousand meters overhead in the dark gray sky, indicating the speed and direction of the wind in which the swirls of air were embedded. A passive microwave instrument read the quantity of water vapor and droplets in the atmosphere between the observatory and the top of the atmosphere—at the moment we looked, the computer blandly reported there was a bit more than a tenth of a millimeter of water in the sky. A radar device read the height and thickness of the clouds and the size of droplets within them and whether they were rising or falling and how fast. All this work—around $5 million a year for Barrow and Atqasuk, including data compilation and analysis—for the purpose of improving how GCMs handle clouds.

There are few eureka moments in this kind of science. Bernie's program took equipment on an icebreaker project and gathered data to test how well models were handling the infrared energy of water vapor in the Arctic on clear days. They found out the models then in use didn't work as well as an alternative scheme. Not the kind of discovery that makes you the talk of a cocktail party, but a bigger one than most scientists in modeling get to make.

Late in his career, this work opened a new theater in Bernie Zak's life. He had started working at Sandia in 1974, when it began to branch out from nuclear weapons development; he studied nuclear winter. In 1991, he won the job of choosing a location for the Arctic portion of the ARM program. Barrow made sense for all the technical reasons and because there was a science-literate community there. But Bernie chose to work in the Arctic because his wife, Nancy, is one-quarter Iñupiaq, although she didn't know it until adulthood. Her father's father had been an engineer on a whaling ship that overwintered in the Canadian Arctic. The son he gained that winter, Nancy's father, he dropped off at an orphanage in the Aleutians. That child spent his life passing as part French, so ashamed of his real heritage he threatened suicide if his own children ever learned he was half Eskimo.

Bernie took on the role of a science grandpa for Barrow. He was a tall, gangly man with a huge smile and twinkling eyes; he wore sharply pressed pants and a hat with button-down ear flaps that looked like it belonged at a late-autumn football game. He helped out at BASC and in the schools. He spent months of the year at a plywood duplex at NARL, employing young Iñupiaq technicians to run the high-tech equipment and launch radiosonde balloons. He even helped hire anthropologists to interview elders, although what they could say about atmospheric physics I never understood. When Bernie talked to schoolchildren for BASC, it was clear he had spent the time to think out how to express the big picture of climate change in a way they could understand. He planned to fly off to a tiny village to give a talk one cold day, but the organizers canceled because only one villager was there to listen; nonetheless, I think he would have gone. He was a favorite of Justine, the daughter of Glenn Sheehan and Anne Jensen, who threw big, sloppy hugs around him.

Bernie was an optimistic man. What he had seen happen over his career helped make him so. When he was an undergraduate, DePaul University in Chicago did not have a computer. He saw one once, but although it filled a large room, it was less powerful than today's pocket calculators. Four decades later, his work with ARM involved banking immense seas of data gathered from one point on the shore of the Arctic Ocean for the purpose of honing computer code on the fastest computers ever built. That was a long way to come, but computers still weren't fast enough for GCMs. The fastest computers would need to be ten times faster, even with big strides in parameterization. The wisest modelers I talked to agreed that decades remained before a climate could really live inside a computer and make firm predictions. But when I brought that up to Bernie—the immense distance ahead, and the small steps all his work would add to that journey—he brought up the Wright Brothers. And we were way ahead of where they were, he said. We were up to the DC-3 in climate modeling, according to Bernie.

"I personally feel really privileged to have even a small part to play in this issue," he said. "I feel it's going to be a major issue for mankind for a long time. If we succeed in not blowing ourselves up, it's going to last thousands of years.

"Even if one is not in a position to answer these questions, it's really interesting to be in the game."

The supercomputers in Fairbanks stood behind a window visible to the public off a hallway at the Arctic Region Supercomputing Center, but they didn't look like much. Big computers needed to be small to minimize the time for electrons traveling at the speed of light to get from one part of the machine to another. There were several standing on a white floor, the size and shape of large appliances, impassive, begrudging even the satisfaction of blinking lights. They might as well have been refrigerators. My hosts were used to giving tours and knew the computers themselves were not the highlight. They next took my son Robin and me (Robin at ten was fascinated by computers) to visit the visualization lab, where Robin put on virtual-reality goggles and drew a picture in three-dimensional space, creating an object that floated in the air and could be viewed from all sides and made to move nearer and farther away. The lab's 3-D printer could turn these virtual drawing-sculptures into three-dimensional objects in the real world, working like a cross between an ink-jet printer and a hot glue gun to build up material to match the image in the computer. Really cool.

A dozen years earlier, Syun Akasofu, the NARL veteran and aurora expert, had helped persuade his friend Ted Stevens, then ranking member of the Defense Appropriations Subcommittee, to insert $25 million into the defense budget to pay for the center's first supercomputer. Ted funded many questionable projects over the years, but none created as much of a stir as this foray into the prerogatives of competitive science funding. *Science, Nature*, and the *Washington Post* all ridiculed the project and the senator's floor speech effusing about how Syun's research would harness the aurora borealis to solve the world's energy problems. In the end, however, the two friends felt vindicated: the supercomputing center in Fairbanks thrived, with successive generations of new computers and a staff of thirty or forty. In 2002, the center installed for testing, free of charge, a Japanese supercomputer that was a scaled-down version of the fastest in the world, the Earth Simulator, thus breaking a long-standing industry barrier to the entry of Japanese high-performance computing in the United States. A variety of computers with different architectures served clients on the campus and all over the country, about a third for Arctic research and the rest for unclassified defense research on oceanography.

Syun Akasofu spoke modestly of his connection to Ted Stevens, saying he didn't have much contact, let alone influence (Ted himself didn't agree). Academics weren't supposed to be politicians. They faced bitter envy and righteous denunciation when they stepped outside the normal channels of competitive, peer-reviewed scientific funding. As a smart politician, Syun deflected attention from his own successes, but at heart I suspected he enjoyed the consternation of the academic powers. The legend of his career that he often repeated was a story of subversively overturning the weight of scientific orthodoxy, persevering in the face of the initial rejection of his theories about the aurora. Colleagues warned his adviser to drop the work or damage his own career (that was Sydney Chapman, the cochair of the International Geophysical Year, who brought Syun to the United States). At first Syun couldn't even get his work published, and he had to wait two decades to be vindicated by satellite imagery. The experience inspired his view, remarkable for a practitioner of hard science, that "what we call truth is not really truth, it's just an idea agreed [upon] by a large number of people." The job of the scientist, he said, was to listen to nature, not other scientists, and to remember that ultimate understanding will never be possible. "A scientific establishment is highly conservative and will attempt to preserve the power of its ruling group against any rebels," he wrote, adding elsewhere, "Since I have an instinctive tendency to avoid prevailing ideas and theories, I am perhaps not a normal scientist."

Syun maintained his political connections with Japan. In more than forty years he never adapted entirely to American culture. America's late-night TV comedians baffled him. Back in Japan, he privately lectured the late Emperor Hirohito and successfully solicited a beer company to fund part of a rocket range near Fairbanks. In the mid-1990s, he began lobbying both countries for an institute in Fairbanks to study Arctic climate change. The International Arctic Research Center (IARC) opened with a shiny $30 million building in 1999, funded jointly by the United States and by Japan's Frontier Research System for Global Change, the same program that owns the Earth Simulator. The center's mission was to bring a large, coordinated team to study the changes in Arctic climate using many tools, but especially the supercomputer a block away on the Fairbanks campus. Ted Stevens pushed through $5 million a year for three years from the National Science Foundation; the NSF refused to pass the money on, instead calling for a proposal, as in the competitive process. Syun Akasofu's group "won," but there was no real competition. In 2003, when that first round of grant funding was running out, at least one senior NSF official expressed disappointment in the level of international cooperation going on at IARC and sent signals it might not be funded again. The Senate Appropriations Committee formally urged the NSF to continue funding. The grant proposal Syun sent in provocatively questioned if observed Arctic climate change could be blamed on human CO_2 emissions or if it was driven more by natural climate cycles. An independent review committee approved the grant. Syun was at the point when scientists usually slow down or retire, but IARC was still a wobbly foal that needed his international political maneuvering to stay on its feet. Richard Glenn, Suki Manabe, and Norbert Untersteiner, the ice scientist, joined the five-member science advisory committee, and IARC's faculty included some top people, but the institute remained a political creature.

One problem was the alliance with Japan. Richard Glenn said Japanese researchers sometimes descended on IARC like science troops rather than individual minds. Suki Manabe ran into the same problem when he returned to Japan to work on the Earth Simulator in 1997. Researchers there stayed with their teachers and institutions permanently; they lacked the initiative to break away and collaborate with a new leader with different ideas, even the founder of the field. After four years he quit and retired to Princeton, believing Japan's scientific culture made it unable to compete. Syun, who coincidentally came to the United States the same year as Suki, 1958, never returned to Japan to study, only to lobby. His iconoclastic style seemed utterly unsuited to that culture.

I spent a week with my computer set up in a pristine cubicle in the silvery IARC building, gingerly knocking on the doors of modelers I wanted to talk to, aware I was breaking trains of thought whose length I could only dream of. The scientists were warm and helpful, but their work was invisible and, except for its broad outlines, impossible for me to understand. Modelers spent days staring at plots of gradations of colors on maps. They thought about sophisticated statistical analyses of environmental data to shape it into forms that would allow them to evaluate comparable model output. They filled pages and white boards with equations. Rather than build another GCM to add to the dozens already running at centers around the world, IARC modelers were comparing and refining the Arctic portion of models built by others. They had spun up a clone of the latest model from the NSF-funded National Center for Atmospheric Research in Boulder. It took a year to get it to work properly on a computer in Fairbanks. Then they began making changes, carefully so as not to get lost, and struggled to see what happened as a result. They built smaller models to understand the changes in the big models: models with versions of ideas simple enough to understand without setting off a hopelessly complex cascade of reactions in a GCM. They tried to find mathematical patterns in nature that matched the patterns produced by models. It was terribly hard work. "It's very frustrating," said a middle-aged modeler. "Am I wasting my life?"

A profound quiet pervaded the building, the sound of an academic institution without students, without many spoken words, with neutral fabrics on the walls of cubicles and little windowed offices like an insurance company's. Everyone here worked first and most intimately with a computer. Discussions occasionally broke out in hallways and in a few seminar rooms, but it was easier to communicate by e-mail, even within the building, even to the next office. Writing e-mail forced clarity of thought, prevented unwanted interruptions, and allowed colleagues to participate in discussions regardless of where they were. Meetings were held on the Web, too. Older scientists said they sometimes missed the days of ad hoc talk with colleagues when new ideas would suddenly bubble up, inefficient conversations that sometimes led somewhere unexpected and important. Those encounters still happened sometimes, perhaps more at IARC than at similar institutions less committed to international cooperation, but the overall sense of the place was of many people alone together. The sound was the same at every hour, during or after work.

A large section of offices established for a program of international researchers to collaborate at IARC remained an immaculate zone of empty desks and cubicles. After the facilities were set up no one came; the researchers saw no need to go to Fairbanks physically. Computers had eliminated the need for bodies. The space itself was similar to the one next door where IARC modelers were working. Standing alone in the spooky vacancy, I felt as if the scientists had disappeared from the physical world and become virtual people.

A few times a week the modelers would convene for seminars. A Russian job candidate gave a talk to show off his stuff, standing before a room of brilliant and famous modelers and Arctic scientists, as stressful a position to be in as I could imagine. He wore a gray pin-striped suit, the only one I had seen in Fairbanks, where any shirt with buttons was formal; he had removed the tie, but it didn't help. The talk began with a blizzard of jargon as he explained a model he had built. Questions fired forth from a Russian scientist in the audience—a tough man of blazing intelligence publishing important new work all the time—and the discussion quickly degenerated into technicality and incomprehension by both men. One question took several minutes and two slides before the two sides understood each other and the speaker could answer. Then a Japanese scientist joined in the attack, openly skeptical, challenging the whole basis of the mathematics in the model. The job candidate was at a loss, either unable to understand the questions or recognizing that his ship was already disappearing beneath the waves and giving himself up to the deep. An American-born scientist, John Walsh, stepped in with softening remarks, defusing the impossible tension in the room, offering an interpretation that bridged the two sides, allowing the seminar to end without further pain. But the guy didn't get the job.

At the next seminar I attended the speaker was Cecilia Bitz of the University of Washington, who had been visiting IARC. She was well respected in her specialty of model comparison of Arctic sea ice and well liked by the staff, but she, too, seemed nervous in front of this crowd. She explained, as everyone present surely knew, that the positive feedback of sea ice reduction and the associated changes in albedo and water vapor were the source of the Arctic amplification of climate change. Although sea ice covered only 4 percent of the globe, it was responsible for 20 to 40 percent of the warming, most of it from the greenhouse effect of water vapor that was released from a sea with less ice. That was a conclusion from models. Then she put up a slide showing a bunch of psychedelically patterned circles, which by now I recognized as output plots of GCMs showing conditions in the Arctic. Cecilia compared present-day sea ice in seventeen models, all different, none correct. "Why on earth do these models have such a big difference?" she said. There were subtle problems as well as the obvious ones. The questions started. And now I saw the same phenomenon as at the Russian's lecture. Despite everyone's being experts and friends, the questioners and the speaker couldn't seem to understand each other; each question required extensive back and forth simply to make apparent to each side what was meant. I was relieved to see I wasn't the only one who had trouble with this material. The work was so abstract that words couldn't handle it. Moreover, scientists who spent all their time looking at computers weren't good at expressing themselves. Again John Walsh stepped in to help, translating between the two sides.

John had spent thirty years teaching general meteorology at the University of Illinois. Many scientists who teach complain about the cost in time and distraction, complaints similar to those of the grant-funded research professors running on the treadmill of constant proposals to make their salaries. To be able to think about work alone, without the pressure of teaching or proposals, was the rare privilege of some scientists at IARC. But teaching could also make a better scientist. John knew how to say things so that people could understand him. He knew how to find the level of the person he was talking to quickly and inconspicuously, deploying a sort of knowledge sonar. He used it on me without my ever hearing a *ping*. Many young scientists, especially those who had never taught, could speak only in technical terms that their expert colleagues could understand or in patronizing terms appropriate for their grandmothers. But the science of climate change demanded that many disciplines work together. Oceanographers needed to talk to atmospheric physicists and chemists and biologists and ice scientists. Talented teachers who were good scientists were invaluable, people such as John Walsh. I heard them called translators, the senior people who could stand up when a seminar or conference was descending into a hopeless tangle and pull everyone back to the broad, important questions that really mattered, using words that everyone could understand. When IARC stole John away from Illinois it was announced as a significant accomplishment.

John looked like a scientist. He had a long, aristocratic face, short hair, and glasses; he would fit right into a film of NASA mission control in the 1960s, with the clean-cut men in short-sleeved button-down shirts leaning over big metal consoles. He had started his career at Dartmouth in New Hampshire and studied sea ice for the passage of an experimental icebreaking oil tanker called the *Manhattan* before the Alaska pipeline was built—many scientists of his generation were drawn to the Arctic by research money spent on oil development in the early 1970s. He was older than many modelers, coming from a time before computers took over communication between people. A workshop conducted on the Web lost something, he said, because you couldn't see the body language of the other person, couldn't go back and forth to quickly negotiate your terms. If you weren't on the same wavelength right away, the thread was lost. Ideas fell into a crack to oblivion.

More and more finished work was going to oblivion, too. Work was published but never cited, never read, and made no contribution to advancing science. The field had exploded with so many journals and so many scientists publishing ideas and findings that it had become impossible for anyone to read it all. Scientists' only hope of keeping up with the journals was to specialize narrowly, yet climate modeling required knowledge of a range of disparate fields. So oblivion yawned wider. Meaning required a context and a human connection, a community to share and appreciate one another's work, to decide what mattered and had to be understood by everyone. But, like e-mail that goes astray in cyberspace, many papers just flowed into databases and libraries to die. E-mail was more efficient than talking, but sometimes people stopped responding because the in-box filled too fast to read through. Communicating electronically, you never knew for sure if what you said had been heard. When I pictured the oblivion of unread papers and unheard ideas, I saw that dark office of empty desks at IARC, built for the scientists who instead vanished into their computers. People did matter, their bodies as well as their minds.

Despite these problems, John was an optimist, like Bernie Zak, the gangly Barrow atmospheric physicist of John's own generation. Climate modeling had come a long way and was making significant progress all the time, as he frequently pointed out. Besides, there was no other tool to predict what would happen as greenhouse gases increased.

"There's a uniqueness to what's going on now, which is what makes it interesting and also what makes it difficult," he said. "If you're talking about a disease, the physician can look at millions of similar cases, whereas the earth probably has never had an increase in CO_2 as fast as we have now."

Even in biology, however, a realistic computer model of the biochemical reactions of a single cell—much less a model of the entire human body—was thought to be ten years and $150 million away. In biology and climate modeling, some hoped for a great insight that would shortcut brute force. Physical predictions were made before computers: Johannes Kepler found the motion of the planets without studying every molecule in the solar system. One such idea under study was to conceptualize the earth as a heat engine and find the thermodynamic entropy of the system. I came across similar thoughts in an online workshop sponsored by the Fairbanks-based Arctic Research Consortium (a nonprofit group funded by the NSF to help coordinate the field) and looked up the author, who turned out to be a young research associate at the University of Colorado named Johnny Lin, a would-be Kepler for Arctic climate science.

Johnny, who studies Arctic sea ice and the atmosphere, had never been to the Arctic, but he had thought about it as an abstract system in creative ways. He started with the implications of one of the oddest findings of the field, that the interacting parts of the system by themselves could generate dramatic changes in climatic conditions without an external causative force. Somehow, the feedback loops and linkages combined and interfered with one another to produce patterns of change in air pressure, temperature, ice movement and abundance, and many other factors. El Niño and La Niña, or the Southern Oscillation, was another, more linear example, a set of interactions in Pacific trade winds, sea level gradients, and temperatures that looped back on itself through its own logic. The Arctic was a similar but more chaotic machine; it had no smooth circle of return to the starting point, like El Niño. Somehow, strange things happened, like a box of bouncing dice that came up all odd numbers, then a little later all primes.

Johnny attacked the Arctic in his computer as a mathematical puzzle. "In a sense I am trying to create theories that may not physically be true but get you the actual insight. Which seems a little strange, but that's the way I'm thinking about it right now. Even when this work is all said and done, it won't say how this system functions, but it will help us understand what happens."

Mathematics might be able to comb chaos from apparent randomness and find levels of complexity with recognizable signatures. "It's probably best understood at this point as a metaphor," Johnny said. The tools of information theory developed in electrical engineering might decode messages that parts of nature were sending to other parts. If you looked at the sea ice as a written page on the ocean, what could you read that it was saying to the atmosphere? "There's times I wonder how fruity it is, but you plug away at it and hope you can get something out of it."

Johnny, like a growing number of other scientists, no longer believed that loading more and more detail into GCMs was leading toward an answer. His mathematical gymnastics were aimed at finding a simpler path. The models were already too complex for a human mind to comprehend. Models were needed to interpret the models. I wondered if piling abstraction on top of abstraction would move reality any closer. Among scientists who spent all their time with computers and none of their time witnessing the Arctic itself, reality was a distant image distorted by thick glass. Nature never seemed farther away than in those quiet offices where computers had silenced human voices. It seemed improbable to me that truth lay in that direction.

During the week I visited, IARC's modelers and some other scientists gathered in a lecture hall to hear a two-part talk ostensibly about modeling the creation of a continental ice sheet during the last glacial period. The Laurentide ice sheet that shaped North American landforms 21,000 years ago was three miles high, more than large enough to affect climate. Gerard Roe of the University of Washington carefully laid out each technical consideration in his model, clearly resolving questions offered by the elder translators in the audience and marching confidently, steadily onward through the lecture as if pushing through a thicket with his solid physique and tough English accent. And then he arrived at the clearing on the other side. Gerard's calculations showed that the presence or absence of the ice sheet depended on changes in energy half the size the best models could hope to resolve. "There is a fundamental knowability limit," he said. "We shouldn't expect to get the right answer."

Having seen the treatment the Russian job candidate received, I expected the audience to explode. The lecture had built a persuasive case that the whole GCM enterprise to which the institute was dedicated wouldn't work—that massive changes could come from inputs much smaller than the models' margins for error. But after a couple of technical questions the scientists filed out quietly; the only person who stayed to talk to Gerard at the front of the room was someone interested in Alaskan glaciers. The message was discouraging, but most modelers had no illusions that they were predicting the future.

I took Gerard to lunch the next day at a Thai restaurant in downtown Fairbanks. He quickly picked up his side of the fight in the battle that had failed to erupt from the IARC audience.

"At some point someone is going to ask, 'What have you got to show for forty years of research?' And I'm afraid the answer will be not that much," he said. "People don't understand the earth, but they want to, so they build a model, and then they have two things they don't understand."

The problem was the same one John Walsh had pointed out—diagnosing a climate with too much CO_2 was like predicting the outcome of a disease based on only one patient. But Gerard suggested that the problem wasn't just an interesting challenge, it was a fundamental collision with the scientific method. A physicist should be able to do an experiment to test an idea and then do it again. "There's no equivalent for looking at a complex climate system," he said. "We only have one planet."

I had been asking modelers what they would do if a model went off in a wildly unexpected direction of sudden cooling or warming. Each said he or she would fix the model so it more closely matched the relatively smooth climate changes we observe in the real world—there was enough room in loosely defined parameterizations to make adjustments that would correct the problem. In fact, that was a common process in the difficult job of building models. But when I asked Gerard that question, he pointed out that the real climate isn't smooth at all: it also contains a record of drastic, sudden changes. Ice cores drilled in Greenland, confirmed by other findings, show many examples, the most severe being the Younger Dryas, a 1,300-year period about 12,000 years ago that saw a cooling of 15 degrees C. That change happened in a decade or less, at both the beginning and the end of the cold snap. Gerard said a model that spiked that way would be assumed to be wrong. In fact, he said, no one can rule out drastic change or even put good bounds on its likelihood. "There used to be crocodiles on Ellesmere Island," he said. "No one has any idea how you get the climate to do that."

Suki Manabe's three-dimensional model built in 1974 showed an increase in global temperature of 3 degrees C for a doubling of CO_2 in the atmosphere. The most accepted figure after another thirty years of work on many GCMs was 3.5 degrees C. By adding more precise parameterizations, finer resolution, and more detailed processes, GCMs might eventually get to the point of being truly predictive of climate in all its complexity. But Suki and the colleague who was carrying on his work at GFDL, Ron Stouffer, didn't think that could happen for three or four decades at least, perhaps much longer—by which time it would be possible to look out the window to get the same information. Indeed, if the question for models was, "Will human CO_2 emissions cause climate change?" we already knew the answer beyond a reasonable doubt. Suki got quite excited about it when I asked him: "It is unthinkable, if you increase the greenhouse gases, that the climate will not be warmer. No one would think that."

You didn't need a model to get this, you just needed a clear mind. The basic physics of how energy reaches the planet tells us the most important driver of climate is the sun, and the second most important is the greenhouse gases. Greenhouse gases were increasing fast, making them the largest factor that was changing. The most reasonable base assumption, therefore, was that the earth would warm. It was certainly possible to argue that other, smaller effects would cancel greenhouse gas warming, but the burden of proof belonged on those who made such claims. Yet even as evidence piled up on the side of the most reasonable assumption, the public debate repeatedly spun off in weird directions, with every "what if" given equal weight. It was as if a murder defendant caught with a bloody weapon in one hand and a written confession in the other were acquitted on the theory that an alien might have beamed him into that position. Perversely, the millions spent on big GCMs to predict the future might simply have delayed recognition of the obvious. Striving in that direction made a statement that we didn't know everything, when the right statement might have been that we knew a lot, or even that we knew enough. Amanda Lynch, the atmospheric scientist who directed Johnny Lin and other modelers at the University of Colorado, asked attendees at an international GCM conference to name the most important question in the field. All present said it was to predict the change in global mean temperature. But besides being fundamentally impossible, Amanda said she didn't think that was even a good question. "I don't think there is anyone who really cares about the global average temperature," she said. "People care what is going to happen where they are." And modeling already had the tools to tell them something about that, at least in a broad, statistical way.

Suki Manabe said the push for predictive modeling came from government, not ordinary people. Funding agencies demanded more realistic models and scientists responded by adding more physical processes, but each new detail brought more possibilities for error. "Uncertainty keeps increasing with the more research money they put in," Suki said. "The uncertainty of the severity of future climate change is increasing over time rather than decreasing. . . . It hasn't gotten any better than when I started forty years ago."

Suki established a philosophy at GFDL of including in models only what he firmly understood and asking the models only questions they could clearly answer. At first, he had no choice. The slow computers available in the 1960s limited modelers to firmly established physical laws and the broadest brush of climate features. The smudging of detail was dictated by calculating limitations, but it may have elicited the best answer nature was ready to give. And it certainly sounded like more fun than what many modelers were doing now. "We avoided getting tangled up with this complicated maze of computer programs that we can't understand," Suki said.

Today GFDL had a huge parallel computer humming away in the basement in a big, dark room accessed through a storeroom where metal shelves were stacked with obsolete dot-matrix printers and the like. The air in the computer room was so clean my allergy-prone sinuses suddenly opened up; I kept asking questions so I wouldn't have to leave. Ron Stouffer said GFDL bought a new $40-million supercomputer every five years or so; the next one would be among the world's top thirty in speed. But the researchers still followed Suki's philosophy of simplicity, clarity, and reasonable questions, and on that basis they were learning a lot.

Suki and Ron built a model in the 1980s that coupled models of the atmosphere, ocean, and land to test how climate change could affect currents in the Atlantic Ocean. The thermohaline circulation (called the THC), driven by differences in ocean temperature and salinity, moves water like a north-south conveyor belt. Warm water rides north, bringing energy from near the equator to the North Atlantic and helping keep European weather mild. Suki and Ron found a relatively modest addition of freshwater to the Arctic and North Atlantic could suddenly stop the THC from flowing. The increased precipitation and melting ice of climate change could produce the freshwater. They had found a climate switch that seemed to explain the sudden and drastic cooling of the Younger Dryas, and that suggested it could happen again.

Since their paper appeared in 1988, a lot of research concentrated on the THC has confirmed the basic finding. A shutdown of the THC would chill the North Atlantic region and heat the tropics, affecting weather all over the earth—for example, a hotter tropical ocean might tend to produce bigger hurricanes. Heat that couldn't move northward in the ocean might instead power a more violent atmosphere. Another 1,300-year chill like the Younger Dryas, arriving with the suddenness of a single decade, could shape human history. And a weaker THC would reduce the ocean's ability to absorb CO_2, further aggravating the carbon cycle's imbalance.

Observations in nature showed that the GFDL model was seeing something real in the world. The THC has varied in strength with changes in climate and CO_2. Preliminary measurements suggested it was weakening now.

CHAPTER SEVEN

■

The Signs

FINDING MY WAY into Barrow's whaling culture was a challenge. National journalism about Barrow's Iñupiat had often been offensive and laughably inaccurate, portraying them as gauche Arctic Arabs, drunk on power and oil wealth. A long article in *Outside* magazine headlined Barrow as "the worst place in the U.S.A." and described the people as "the Beverly Eskimos," both menacing and lazy. From the perspective of many people in Barrow, outside journalists represented a hostile worldview, one that assumed that only certain people deserved wealth, that hunters were evil Bambi killers, and that whales were singing saints of the deep. Iñupiat leaders could not change those presuppositions, but they could deny access to their whaling community, and they usually did. In the fall of 2001 I went to meet Richard Glenn for breakfast at the Hotel Captain Cook in downtown Anchorage and found several top ASRC executives at the table, including Oliver Leavitt and Jacob Adams, and it slowly dawned on me that I was being sized up.

I had some credentials. I was an Alaskan involved in my own community. My father had worked for Governor Bill Egan, who remained a beloved figure in Bush Alaska twenty years after his death. And Richard liked an article I was writing about whaling and traditional knowledge. But the ASRC executives had other considerations on their minds, too. Congress would soon be taking up ANWR oil exploration: Oliver tested me on that issue, but I just wrote down what he said and asked another question. And they were skeptical of climate change, or at least worried it could be exaggerated. The changes they had seen were real, but they were conservative interpreters of nature and the new Arctic science was still primitive. Oliver and Jacob said the beach had been fifty to a hundred yards farther out when they were young, but offshore gravel mining in the 1950s and 1970s could have accelerated erosion. The western Arctic village of Shishmaref was washing away, but it was built on a barrier island that could not be expected to last forever. The sea ice was thinner and less reliable, but that could be caused partly by ocean and atmospheric circulation that might not be directly related to warming. The weather was warmer, but maybe it would get colder again. No two winters were ever the same. Weather got warmer and colder. How long would a run of warm weather have to last before it became a change in climate? "People talk about global warming," Oliver said. "You hear it long enough, there must be global warming."

My son Robin once asked me, "How do we know if our prayers are answered?" The deus ex machina of climate change presents the same conundrum: How could we recognize in everyday life a difference that transcends days and years? Even scientists who accepted the physical inevitability of climate change disagreed on whether it was manifested in the Arctic's present warming. No one predicted the change would come so fast or dramatically, so the hope persisted that this would turn out to be just a long warm snap. But that could be wishful thinking, as deceptive as the suggestibility that Oliver was worried about. For lovers of winter like me, the idea that real winter might never return was almost unbearable. Sitting with the ASRC executives, I wrote in my notebook and kept my mouth shut. The men around the table talked with half an eye on me, joking about the dubious evidence of climate change. I didn't really believe them. They sounded like they were trying to convince themselves, as bluff as kids passing a graveyard.

Science has struggled to separate the strand of climate change from the weave of the environment. If natural changes in weather and climate were random, the problem would be much easier, as statistical analysis is well suited for pulling a rising trend line from a random background. But the climate record is full of sudden break points and cycles of rising and falling temperature on many frequencies. The glacial cycles of the last 2.5 million years have oscillated on switching frequencies of 21,000, 41,000, and 100,000 years, in rough synchronization but not in proportion with oddities of the earth's path through space in relation to the sun. The Medieval Warm Period and the Little Ice Age, arriving within the current interglacial warmth of those larger cycles, may have been waves in their own 1,500-year cycle. The end of the Little Ice Age, around 1860, initiated climate warming and sped the rise of sea level long before carbon emissions from fossil fuels became significant. Other postulated oscillations repeated at frequencies of seventy years, fifty years, twenty years, eleven years, two years, and other periods. Each had its own acronym and scientific literature in support and opposition. They could all be right. The climate may reverberate like a pipe organ. From a human perspective, the experience of one-way, long-term climate change could be like driving into mountain foothills, the overall rise obscured by the ups and downs at our own scale. Maybe it is impossible for human beings, with our short life spans and even shorter memories, to be aware of a trend that passes through such long cycles. Many scientists certainly think so, and the Eskimo executives I met at that hotel restaurant did, too, at least that day.

Igor Polyakov, a Russian oceanographer working at IARC, first found the trace of a sixty-to-eighty-year cycle of the Arctic Ocean's sea ice in a model he built on his computer workstation, a finding he published in 2000. After seeing the pattern in the computer, Igor and his colleagues confirmed it in Arctic weather data gathered near the ocean's edge around the globe. Temperatures reached a peak in the late 1930s and early 1940s, dropped to a low in the 1960s and 1970s, and looked to be ready to peak again.

Igor named this climate signal the Low Frequency Oscillation, or LFO. He theorized that the Arctic Ocean could be acting like a clock regulator in response to climate kicks from the North Atlantic. Using sophisticated statistical techniques, Igor and his colleagues subtracted the LFO from the overall warming trend in the Arctic going back to 1875 and found that it erased the Arctic amplification found in theory and in models. But for the cycle, warming in the Arctic was the same as warming elsewhere on the globe. A controversial result, to say the least. The Arctic amplification—the Arctic's leading role in climate—was a major justification for funding of the region's research.

But the LFO didn't catch on. The record of observations encompassed less than two full cycles, and that included the sketchy weather station coverage of a century ago. Many wanted to see three or four cycles at least before buying into a repeating system. Roger Colony, who directed the IARC/Frontier program where Igor Polyakov worked, set out to find other evidence of sea ice extent that could back up the LFO cycle. Ship captains hunting whales and seals recorded the ice edge as long as four hundred years ago in the Barents Sea, north of Scandinavia and western Russia. Roger's colleagues at the University of Hamburg translated ships' handwritten logs from a mixture of Latin and Frisian to recover that data, providing some confirmation for the LFO.

Roger also collaborated with scientists from Moscow State University who were studying the Russian Chronicles, descriptions and drawings of crops, weather, and local events kept by monks in many Russian towns beginning in A.D. 860. They translated the chronicles' ornate, antique Russian and categorized the descriptions into seven generalizations of severe weather; for example, the 1251 description "the early-in-autumn frost abolished all the harvest, destroyed the abundance" became "cold autumn." The entry for 1330 contained a scary but eloquent drawing of dry fields, a burned forest, and a fish flopping from a dried-up river; it went down as "drought." When the categories were tabulated, they produced a pattern of extreme weather events that correlated well with tree ring data. "I've had a hard time getting anyone interested in it," Roger said. Physicists wanted numbers, not descriptions.

Although Arctic weather records didn't reach far enough into the past to confirm long climate cycles, they did provide a rich diet for sophisticated statistical analysis by atmospheric scientists. The most famous of these calculations were accomplished by John M. Wallace, known as Mike, and David Thompson, who was Mike's student at the University of Washington when they found a pattern they dubbed the Arctic Oscillation. The AO is a mathematical creature; it is best defined in language as an index that reflects a seesaw reversal of atmospheric high pressure in the Arctic and mid-latitude Atlantic. The low index matches weak westerly winds and a meandering jet stream blocking the eastern flow of weather and allowing cold air to drop down from the poles; a high index equates to stronger winds and warmer temperatures as air has more of a chance to mix west to east. Since the late 1960s, the high index was dominant, suggesting that much of the observed warming could have been caused by these patterns of circulation. The idea caught on quickly. Scientists found correlations to other Arctic phenomena that lined up with the high AO index, including thinner sea ice. Journalists were excited by the contrarian idea that atmospheric circulation rather than global warming could be responsible for the changes seen in the Arctic, and the AO swept into some popular writing as a debunker of climate change.

In the fall of 2002, Glenn Sheehan invited Mike Wallace to Barrow to speak at the Iñupiat Heritage Center, part of Glenn's guerrilla campaign for the new building—he was using BASC's educational lecture series to bring in academic stars so he could show them the town's support for science and its poor scientific facilities. Mike was almost giddy to be in Barrow: the Arctic was the arena of his academic success and his childhood dreams, but this was the first time he had set foot there. He was charming. A slender, slightly rumpled man, the stereotypical academic, he was so enthusiastic about his science that his eyes glowed and his words rushed out to share what he had found. Before an audience of around seventy, mostly ordinary community members, he explained how the high AO index over recent decades, combined with the El Niño–Southern Oscillation, explained most of the winter warming and ice reductions in Barrow. Besides the warming, wind patterns had floated ice away from Barrow.

Judging from the question and answer period, the audience was struggling. Was there no global warming, or was global warming driving the AO? Mike admitted he didn't know; what drove the AO was a focus of research. An Iñupiaq man rose and told how he had experienced warmer winters, weaker sea ice, and stronger storms since the 1950s and 1960s. "The changes on the temperatures are extremely different," he said. "You could walk across the lagoon by the second weekend of September, but that doesn't occur anymore." Mike accepted the comment and tried to relate it back to his theory, explaining that the observations could be consistent with both the AO and global warming. The next comment mentioned the more frequent breaking away of sea ice and the extraordinary shrinkage of glaciers in Southcentral Alaska. Mike returned to the relationship of the AO index to ice velocity—the idea that ice had simply floated away. After the talk, at the back of the room, Mike's wife, Susan, critiqued his talk harshly. She said he needed simpler terms, better definitions.

But it wasn't Mike's fault. Amanda Lynch, the University of Colorado atmospheric scientist and modeler, said many scientists did not understand the AO either. She could tell by the way they misrepresented its implications to boost their grant proposals and spice up their comments to the press. Within the narrower field of atmospheric science, with its challenging mathematics, the AO remained a controversial hypothesis. Amanda blamed the undeserved acceptance and popularity on a catchy name: the Arctic Oscillation sounded like a grand ordering principle for the North, a single pendulum controlling the weather, perhaps another El Niño. "I don't think the Arctic Oscillation is a real [physical] phenomenon," she said.

Mike agreed that many people misinterpreted the idea. The pattern was never intended to explain away climate change. Instead, it represented a way in which the climate had changed as the system flipped from one mode to another. And he agreed that the name was a problem. When he and David Thompson wrote the paper that first announced their finding, they were looking for a helpful hook for a highly abstract concept. The North Atlantic Oscillation, a similar concept, had been around for a long time; by calling their own idea the Arctic Oscillation, they hoped simply to move people's minds north. But, in fact, the phenomenon was not an oscillation at all, it was simply a set of two possible modes organizing the atmosphere, the high and low index, whose reversal didn't necessarily have regular timing. In their later papers, they started calling it the Northern Hemisphere Annular Mode. A science writer who had helped popularize the Arctic Oscillation wrote Mike a note chastising him for the name change, but he needn't have worried, as the new name showed no evidence of catching on. Everyone still said Arctic Oscillation. "If I had it to do over again, I think it would have been best to get the science right, and then worry about how to make it popularly appealing afterwards," Mike said. "We get into these unfulfilled expectations."

Fundamentally mathematical ideas cannot be popularized without losing much of their meaning, but more accessible climate cycle offerings were available for consumption, too. In 1996, Steven Hare, then a graduate student at the University of Washington, compared salmon catch records from Alaska with those from Washington and Oregon and found that when fishing was good in one region, it was poor in the other. These conditions lasted for a quarter of a century and then reversed, and they matched up with a previously unknown pattern of ocean and air temperatures, stream flows, and sea ice extent. Hare named the phenomenon the Pacific Decadal Oscillation. In the major paper announcing it, University of Washington scientists called for fishery managers to take note. But the discovery was of much broader interest. The PDO linked with a fifty-year cycle of fisheries, bird and zooplankton numbers, water temperatures, air pressure, atmospheric carbon dioxide concentration, and currents that manifested across the Pacific from Alaska to Japan to Peru, reversing roughly every twenty-five years—at the top of the century, the quarter centuries, and the half century. This cycle, exemplified by a seesawing in the abundance of anchovies and sardines, allegedly explained even the fishery crash that inspired John Steinbeck's *Cannery Row* in 1945. Needing a name for it, and mindful of the success of such names as El Niño and the Arctic Oscillation, the authors of a paper in *Science* suggested "El Viejo," the old man, and "La Vieja," the old woman, for the two sides of the cycle.

No one had explained what caused El Viejo (or whatever it would end up being called), but it demonstrated the exquisite sensitivity of ecosystems to small changes in climate. Biological systems may be better than computers or mathematicians at finding climate patterns, because they can amplify a weak climate signal into a cascade of more noticeable effects. In fact, such systems might even feed back into climate like an amplifier in a sound system, blowing up a small climate change into a large one. That idea could help explain, for example, how a small reduction in the amount of sunlight reaching the earth due to a 100,000-year eccentricity in orbit gave rise to a big 100,000-year cycle of ice age periods. The cooling from the sun was too weak by itself to freeze the earth, but it could have driven biological changes that affected the atmosphere's CO_2, and that in turn could have cooled the climate enough for an ice age. British scientists found that contemporary ecosystems can amplify weak, hard-to-find climate signals even when the weak signal lies underneath a stronger one, like a bell that vibrates to a low, resonant tone amid a blaring symphony.

Nature could find variations; nature could also find trends. The temperature of permafrost well below the surface responds to changes in the atmosphere very slowly. It takes a long time for warmth to work its way a hundred meters or more down through frozen ground. Smaller changes merged into larger trends as the mass of earth absorbed the total energy it received. For the scientists who measured permafrost temperatures, raw data was already filtered, climate spikes and cycles equalized like lumps of flour through a sifter. What remained was a smooth record, an average temperature over periods of thousands of years. The length of the average depended on the depth chosen for the measurements; Tom Osterkamp and Vladimir Romanovsky of the Geophysical Institute at the University of Alaska worked with averages up to 1,000 years long, ranging back to the end of the last glacial period.

Besides the work begun in Lieutenant Ray's pit in 1881, permafrost measurements were made on the North Slope starting in the 1940s. Osterkamp began studying drill holes specifically to look at warming weather in the early 1980s. This was low-tech science: a hole an inch and a half wide made with well-drilling equipment, lined with PVC pipe, filled with antifreeze, and wired with temperature sensors and data loggers to be downloaded once a year. Many experiments went on at once. As Vladimir said, the depth of the hole was determined by the quantity of the funding. Shallower holes helped look at changes over shorter time spans—as short as a few years. The overall story was of long, slow climate cycles followed by a sharp upward jump late in the twentieth century. Bringing the story to the present, permafrost all over Interior Alaska that had been solid throughout historic times had melted entirely, collapsing roads and hillsides and turning forests into muddy holes. Even under Arctic tundra, where permafrost was deep and unlikely to melt soon, temperatures were nearing the freezing point, Vladimir said.

The layering and chemistry of ancient ice in deep bore holes in Antarctica and Greenland yielded climate records hundreds of millennia long. Glaciers draped over the mountains of Alaska recorded current climate over decades. Mountain glaciers are flows in the earth's water cycle. Water that evaporates from the ocean falls as snow high in the mountains and slides back down as ice. At the top of a glacier, snow accumulates faster than it melts. The glacier ends at a lower, warmer elevation—which could be sea level or some seemingly arbitrary point in a mountain valley—where the rate of melting overcomes the flow of new ice. Glaciers grow with cooler temperatures or increased snowfall; they shrink in warmer or drier conditions. There are few glaciers in Alaska's Arctic because the climate is too dry, but many of the world's largest, most active mountain glaciers lie on Alaska's southern coast.

It had been obvious since John Muir paddled into Glacier Bay in 1879 that these glaciers were shrinking fast, but figuring out how fast they were shrinking and if the rate was increasing took repeated measurements of the glaciers' total mass. Measuring glaciers in three dimensions was hard. Alaska had 160,000 glaciers; consistent measurements had been accomplished on only one, a small one that was easy to get to. In 1992, Keith Echelmeyer and his colleagues at the Geophysical Institute figured out a quick and cheap way of doing many more measurements, using Keith's three-seat airplane. In a decade they calculated the mass of sixty-seven glaciers that accounted for more than 20 percent of the total ice area, then measured twenty-eight of them again. It was a shoestring operation: a computer installed in one seat of Keith's PA-12, an altimeter capable of measuring vertical distances with a laser, and a precise GPS system, all linked so they could continuously record a glacier's vertical profile while flying over it. Camping out and flying glaciers each summer, Keith took measurements, then compared them with topographic maps made in the 1950s. The difference between the two was the glacial mass gone in half a century, a dramatic thinning of about half a meter a year over the entire period. Returning to the same glaciers six years later, the scientists found that the rate of thinning had more than tripled, despite increased precipitation. At that rate, Alaska's glaciers alone were producing enough water to raise global sea level by a quarter of a millimeter a year.

The increased rate of change in the glaciers could be caused by a natural climate variation, but it seemed improbable to me that the extreme acceleration of the melting of ancient ice would coincide by chance with unprecedented increases in CO_2 emissions. When combined with the permafrost record and many other environmental trends that pointed in the same direction—even the eighty-four-year record of a statewide betting pool on spring breakup on the Tanana River south of Fairbanks—the case was strong for warming beyond natural variation. Natural cycles might exaggerate the upward trend sometimes and hide it at other times, but the peaks and troughs stepped progressively higher like sawteeth on an incline. The burden of proof had shifted to skeptics, in my judgment. Papers by teams of eminent Arctic scientists gathered up the changes, piling up circumstantial evidence until, as a reader, I was tempted to say, "OK, OK, I believe you, move on." Nonetheless, with the conservatism of science, authors hedged and called for more research.

The other reaction I had reading these compilations of evidence was how similar they were to the testimony of Thomas Itta, Sr., Kenny Toovak, and Arnold Brower, Sr., and of elders in Interior Alaska, and of non-Native people who spend their lives outdoors in Alaska. Those observers didn't need to remember the details of the weather a century ago to say something about climate change. They knew their world. When Thomas saw tundra permafrost melt so that it changed the shapes of rivers and streams in unfamiliar ways, he was aware of unprecedented climate change. But he didn't draw a conclusion from that alone. He and the other elders looked at many changes in nature, some unequivocal and some suggestive, discussed among themselves what they had seen, and reached a consensus based on broad evidence. Over long time spans, natural systems synthesized change into recognizable signals; people who lived within those natural systems synthesized the signals they perceived into an understanding of the whole. After all, the human mind, with its capacity for subtle perception, communication, comparison, pattern recognition, and intuition, is the most complex natural system on earth. Why not look there for a summation of the climate's variation and trend?

Richard and Arlene Glenn spent a Sunday in the fall of 2001 with one of Richard's cousins in a tidy, modern house in a new part of Barrow, where they feasted on steak and chicken and worked the day's crossword puzzle from the *New York Times Magazine* while the girls watched a Disney movie on video. This was before our whaling adventures and interviews together, when Arlene was still working at the Iñupiat History, Language, and Culture Commission. This bright, lively couple, with their deep knowledge and diverse perspectives, knew as much as anyone about gathering traditional wisdom and applying it to scientific questions. And they talked about it with the enthusiasm of college students.

At the commission, Arlene had helped compile a great store of information passed on by elders. Each year, the commission sponsored a conference on a different aspect of Iñupiat tradition—whaling, traditional clothing, weather—where elders would tell stories, describe skills that had passed out of use, or dredge spiritual waters clouded by time. A man told how he went adrift with his crew. Taking soundings and finding no bottom, he realized he was far out to sea and found his way home after a long odyssey of many adventures. Advice was given on how to climb out of the ocean when you have fallen off the ice: wait until you are calm, then lie flat and roll. And old spiritual teaching was repeated: that people have spirits that never die, and that animals do, too, and that an animal will be infinitely grateful to a person who severs its head after death, freeing the spirit from its body and from its mourning. The commission recorded these conversations on tape and was gradually transcribing and translating them so that they could be published with the original Iñupiaq and English side-by-side for use in the schools.

Arlene said the conferences were important because traditional learning wasn't happening as much as it should: "We've got more and more things that are distracting, more modern things."

Richard conceded there were distractions but questioned whether conferences were necessary. "Traditional knowledge is moving family to family in a traditional way," he said. The books were helpful for outsiders and provided a way of saving the oldest knowledge undiluted, but they weren't good for passing on the Iñupiaq way of life. "What is happening at the elders' conferences, it's like a spare tire. No one goes to those books to learn these things. You learn from your father."

Richard started talking about learning the Iñupiaq way—how to navigate, how to hunt, how to stay alive. A mentor let you try and fail, broke down your pride to instill humility before nature's power, and put you where you could get a feel for the world and how it works. The teacher was a guide; nature was the real teacher. Now Richard himself had responsibility for teaching younger members of Savik Crew such as Eben and Benny, and he was learning not to be too soft, not to tell too much. He couldn't do it with a lecture or a book. "It's use it or lose it," he said. "The stakes are high. People get killed."

Arlene agreed: you have to do and touch to learn Eskimo ways.

Richard told how he steered the boat onto the back of a whale the first time, hearing in his memory the voice of the elder who had told him not to be afraid, not to pull back. He said, "If you're not all at peace together in the boat, you might as well stay home. You've got to work together."

Arlene said, "That's what they always say when you're sewing the skin to make the boat. You've got to be at peace with the people around you and not raise your voice."

Richard and Arlene both had seen how an ignorant science researcher would go to an elder and say, "Tell me what you know." Some scientists approached Native experts as if they were banks of raw data, free for the taking. Native professionals could behave similarly when they entered into contracts with scientists to supply the traditional-knowledge component of a project, fulfilling a perceived obligation from science funders—it could seem like a sort of racial tax. Richard said race was irrelevant; it was the relationship that mattered. Two experts from different cultures could learn from each other, or a young person could learn from a teacher. But to attempt to atomize an elder's knowledge into data points would destroy it. The knowledge, the person, and the place were inseparable.

Richard asked Arlene how many tapes were in the Heritage Center. She said seventeen hundred. He said, "That's all they are is tapes. The person who holds them is a hoarder of tapes."

That next winter was tough on Arlene. Her father, Frankie Akpik, died that winter. And, at thirty-nine, she was carrying her fourth child. She quit working at the commission and stayed home, preparing to bring a new girl into the world while her oldest, Patuk, finished her senior year in high school and applied to colleges. Patuk had gone to the Alaska Native boarding school in Sitka for a couple of years, at her own initiative, but came home for senior year because she "wanted to graduate a whaler"—a Barrow High School Whaler and a real whaler. Like other young adults before her, she had found her interest in Iñupiaq culture strengthened by her time away. She got involved in Eskimo dance and traditional sports, which included events such as high kicking, knuckle walking, and other tests of strength, agility, and endurance.

Arlene missed work. In the spring I was the first to show her a new book produced by the North Slope Borough School District designed to help pass on the traditional knowledge of whaling. Arlene's former coworker at the commission, Jana Harcharek, had created the book to address a recurring conflict: whaling season came during the school year, usually right around final exams. When teens went whaling with their crews, their grades suffered; some kids even failed and dropped out. Jana, a whaler herself and a bright and commanding woman, put a lot of whaling information in the book, but the heart of it was a list of dozens of educational standards established with the help of whaling captains. For example, "Knows how to determine where the shorefast ice is grounded at vital spots" and "Is able to describe the eight winds" and "Is able to identify bowhead whale morphological types." Each captain would give marks on the level of mastery of each skill for each student in his crew. Students would get classroom credit based on arrangements made with their teachers. Jana's son, Nagruk, the guinea pig in the project, got biology credit for making notes on the dissection of a whale, got math credit for recording whale-catch positions using latitude and longitude, and got English composition credit for writing up stories he heard in camp. The Iñupiat had often adapted outside technology to serve Iñupiaq purposes; Jana was adapting formal schooling to serve traditional knowledge.

"All the years I worked with the elders, they stressed to me that, yeah, it's important to record what we know about our environment and way of life, but it's more important that we find ways to pass that on," Jana said. The cabinets of tapes were pure but dead. The stream of culture flowed on in a modern world with each young person concocting his or her own mix of new and old ways. Jana said, "You take the best of both. You can't make it without one or the other."

The language was changing. Sharp and guttural Iñupiaq sounds that didn't occur in English were washing out; the endings were compressing. The topics of traditional knowledge were changing. More than a century ago, the lance had been set aside for the shoulder gun to kill most whales. Two-way radios and snowmachines had been in use for a few decades. In the 1990s, GPS receivers simplified navigation and made knowledge of water depth and snow patterns on the tundra less important—it was no longer necessary to find location by sounding the bottom or to use the wind as a compass. I never saw any Natives using manuals to operate this equipment: those skills were passed on in the traditional way, too.

The Iñupiat developed their own ways of using traditional knowledge in science and policy at the borough Wildlife Department, beginning with the bowhead population controversy. Harry Brower, Jr., and Taqulik Hepa studied whale, caribou, and duck numbers and behavior using this method. First they would meet with the village as a group, explain the purpose and benefits of the study, and ask villagers what questions they would like to see addressed. They let the discussion proceed at length and without direction until the village arrived at a consensus on the terms of its support for the project. Next, the researchers would interview everyone in the village, in their homes or outdoors, in the presence of the things they were talking about. Experienced hunters often knew more than the biologists interviewing them about animal behavior and habitat areas, and the biologists had information useful to the hunters as well. Once they settled on common terms, the conversation would often catch fire. Finally, the written report came back to the village for review, revision, and approval by leaders, elders, and the community as a whole. The borough then used the reports to help guide fish and wildlife managers, to comment on oil and gas development proposals, and to educate agencies that lacked the money or know-how to gather such information.

Confidentiality was assured from the outset. When I asked Taqulik why, she drew back in surprise. "That's private information," she said. Why was it private? Some things just are. No one could tell me exactly why, although I heard many theories. Competition wasn't the answer—at least not the major answer—because hunters all wanted to pass on their knowledge to the next generation and often shared much of their catch. Glenn Sheehan explained the need for confidentiality by imagining the response an anthropologist would receive on approaching a highly successful member of the dominant society—say, a Wall Street financier—and asking exactly what he earned, how he earned it, and what he had learned about the marketplace. But that didn't quite get it either, because the financier's reticence would be about his social status—his prestige, competitive skills, and worth.

Firsthand knowledge about nature isn't like that, but it can be personal and even intimate. It can be the irreducible imprint of experience, inaccessible to rational dissection, the sensual contact of our physical selves with the real world. The content of a long, silent day spent with a mentor out under the sky, the first day of understanding how a bird moves or how to ride over a wave—that would not be a thing to spill out promiscuously. Some knowledge can only be diminished by being disclosed. A love affair should not be reduced to a list of meetings and acts. An afternoon learning to cook at the elbow of your grandmother—add a little bit of this, heat until it looks like that—should not be reduced to a set of quantities and handed out on recipe cards. Such information is private for cultural reasons, for emotional reasons, but also, I think, for a reason that is more fundamental: this kind of knowledge is what intuition is made of. Not data files that look the same on any computing platform but essential parts of ourselves.

The word *intuition* could get you into trouble. Oliver Leavitt went out of his way to say intuition had nothing to do with how he handled himself on the ice. His skills were based on experience. I think he was responding to a pseudospiritual use of the word. Some Alaska Natives believed that indigenous people were born with environmental knowledge; that they were essentially better than whites down in their bones. Scientists and practical people like Oliver naturally steered clear of such beliefs, which couldn't have a physical basis and were contradicted every time a rural white trapper knew more about the environment than a TV-bound urban Alaska Native. Still, traditional knowledge gained a reputation as being a sort of Native collective unconscious.

That kind of talk led some scientists to think of the movement to respect local knowledge as a politically correct shakedown. People on both sides of the cultural divide complained of projects that included traditional ecological knowledge, or "TEK," like a form of administrative overhead, wasting the scientists' money and the time and goodwill of local people who were sick of dealing with them. Federal science agencies established an interdepartmental policy supporting traditional knowledge, so scientists included it in their proposals to help their chances of winning grants. Using TEK became an academic fad. But Tom Pyle, head of the National Science Foundation's Arctic Program, said, "Other than anecdotal stories, I don't know if we could say we've benefited from it." Each side thought it was doing the other a favor.

But there were successes, especially among scientists who developed personal relationships with individual Natives. Henry Huntington and Dave Norton had both lived in Barrow, and both were Arctic generalists now writing and carrying out NSF grants from home offices in Anchorage and Fairbanks, respectively. Unlike many narrowly focused researchers, they took a broad interest in the world and were capable of talking over coffee about what knowledge was, where it came from, and how it related to its cultural context. With Hajo Eicken and Lew Shapiro of the University of Alaska Fairbanks, and Craig George and Harry Brower, Jr., in Barrow, they embarked on a project to bring together the scientific and Iñupiaq perspectives on shorefast ice, to adopt the Iñupiat's efficient ice-engineering language into the scientific lexicon, and to use the Iñupiat's broad, systematic understanding of ice to enrich the snapshot precision and discrete scales employed by ice science. With success, the project could make whaling safer by improving everyone's ability to predict ice behavior.

For three days in November 2000, the team convened the Barrow Symposium on Sea Ice with five elders and twenty-five scientists facing one another across a quadrangle of chairs arranged around the wall of a room at the Heritage Center. Kenny Toovak, Arnold Brower, Sr., and Warren Matumeak were there, as well as Jim Maslanik and Russ Page (from the weather service ice desk in Anchorage). The experts worked through a series of case studies of extreme ice events, including the big, scary ivu of 1957 and the break-off of 1997, with scientists presenting remote sensing and wide-scale weather data, and elders explaining what had happened at the ice level and why. Some scientists lapsed into technical jargon that neither the elders nor scientists in other specialties could understand, but generally the discussion kept its feet on the ground.

The elders made the most memorable comments: they knew how to tell stories. The paper that came out of the symposium was clear and elegantly written, too—ideas leading organically to a conclusion in a refreshing change from the mechanistic writing of most science papers, which lurch forward in series of short, declarative sentences like Frankenstein monsters of prose. Henry Huntington, the lead author, had been an English major as an undergraduate before he entered Arctic studies. He wrote: "One result of the symposium has been our growing conviction that progress toward genuine integration and synthesis benefited more from the Iñupiat hunters' integrative approach to sea ice observation at the symposium than it did from scientists' input." The elders talked about the shape of the ocean bottom, the water and currents, the underside of the ice, and the atmosphere, and how each factor varied through the year to build the structure of shorefast ice they used for whaling in the spring. The glaciologists talked about the ice as a snapshot of a two-dimensional sheet, bringing in the dimensions of time and vertical space as an afterthought. "Iñupiat traditional ecological knowledge articulated at the symposium was essential in persuading ice scientists to ask new questions, to deploy specific instruments at defined locations in future fieldwork, and to think integratively," Henry wrote.

Russ Page, the ice forecaster, said he learned a lot from Arnold Brower, Sr.—for example, that he had erred the year before by putting out a warning for whalers to get off the ice. Arnold explained that the warning should go through the whaling captains, who were the experts on the scene, rather than being broadcast widely and creating panic. Russ adjusted his practices. He took his responsibility very seriously, perhaps too seriously. At whaling time and crab fishing time in the Bering Sea, life became terribly intense in his tiny cubicle in Anchorage with his computer screens, charts, and records. He called these times "my killing season."

Shorefast ice forecasting was new, and Russell's tools were coarse. The predictions of the four computer models that were supposed to guide him often splayed on diverging hypothetical paths. He had developed his own sense of how ice patterns evolve by looking at maps of ice distribution that covered many years. In his forecasts he would mention which years appeared most similar to the current situation. Arnold Brower Sr. had given him confidence he was on the right track.

"He was the one who thought a lot of my ideas were correct," Russ said. "He was seeing the same things, because of his life span, and because he's sharp. He's stronger at eighty years old than I am.

"A lot of what I was seeing from space, I don't know what I'm looking at. It was the first time we could sit down and talk about the elephant from different perspectives."

Kenny Toovak pointed out how Russ could make his predictions more effective by using simpler language. The glossary of ice terminology adopted by the National Weather Service came from shipping. Mariners simply wanted to avoid ice. People working on the ice needed much more information. Iñupiaq seemed the perfect solution. In English, there wasn't a convenient way to warn of a situation in which large colliding masses of ice would ride up over each other or cause the sudden creation of pressure ridges; in Iñupiaq you could just say "ivu." Russ took the suggestion, and the word was added to the official weather service lexicon. There were many other such Iñupiaq terms that could express complex ideas about ice in a single word, and the science team on the project wanted to gather them and adopt them into Western science.

But that never happened, because the project was canceled. Dave Norton wrote the grant proposal for the second phase, which was to pursue the integrative linkage of ice science and traditional knowledge at four sites around the Arctic, including Barrow. The plan was to allow the participants to help define the study and its methods, making the Eskimos research partners rather than subjects. But the anonymous peer reviewers who critiqued the proposal panned the idea. Their comments made clear that the Natives were to be studied, not to direct the study. Anthropologists, not ice scientists, should interview the Eskimos, and the questions and appropriate answers should be standardized by the researchers before going into the field. One reviewer wrote: "Some semblance of a standardized approach would inspire some degree of confidence that solid results would be produced. The only assessment one can make concerning this plan is that it is 'bad science.' "

The same reviewer suggested that Barrow's traditional knowledge might have lost value because it was no longer pure. Barrow's Natives might know too much about science. "Because the 'intensive study area' does have a long history of collaboration with scientific research, one wonders if a feedback-loop concerning so-called traditional knowledge may have developed," the reviewer wrote. "Researchers may have contaminated the knowledge that the Iñupiat traditionally pass down via verbal communication and demonstration."

That was undoubtedly true, but "contaminated" was a funny word for it. At the butchering of a whale in 1990, Henry Huntington fell into conversation about the animal's anatomy with whaler Silas Negovanna, who explained in detail the difference in the shape of the pelvic bone between males and females. "I was in awe. I said, 'How did you learn all this?' And he said, 'A biologist told me.' To him, it was just knowledge that he had."

Science wants to know where information comes from, testing the strength of each brick going into the construction of its theories about the world. But such confidence alone does not ensure the stability of the resulting edifice. As Syun Akasofu wrote, scientific theories are made to be overturned. They manifest culture as surely as traditional knowledge does. He said a theory can never really comprehend the true complexity of nature, only provide useful ways of organizing it until a better theory comes along. In those terms, the Iñupiat, with their indiscriminate assimilation of whatever information came to hand, had produced a superior theory of how their environment worked. As Henry said, any scientist given the choice of a guide out on the ice would choose an Iñupiaq elder over an ice scientist. "Who cares where the information came from?" he said.

Mr. Ramsay, the academic and hated patriarch in Virginia Woolf's 1927 novel *To the Lighthouse*, imagines himself as a brave polar explorer marching to his demise as he attempts to advance human knowledge by one incremental step. The entire heroic journey is only the chain of reasoning inside his own head:

> If thought is like the keyboard of a piano, divided into so many notes, or like the alphabet is ranged in twenty-six letters all in order, then his splendid mind had no sort of difficulty in running over those letters one by one, firmly and accurately, until it had reached, say, the letter Q. He reached Q. Very few people in the whole of England ever reach Q. . . . But after Q? What comes next? After Q there are a number of letters the last of which is scarcely visible to mortal eyes, but glimmers red in the distance. Z is only reached once by one man in a generation. Still, if he could reach R it would be something. Here at least was Q. He dug his heels in at Q. Q he was sure of. Q he could demonstrate. If Q then is Q—R—Here he knocked his pipe out, with two or three resonant taps on the handle of the urn, and proceeded. "Then R . . ." He braced himself. He clenched himself.

A funny caricature, but also accurate. Scientific advance once was thought of as a march toward Z, but the twentieth century spoiled the sequence. Instead of moving to the next letter, ideas kept cropping up that required a whole different alphabet. The universe turned out to be fundamentally unknowable in some of the areas we most wanted to learn about. Z faded out to infinity.

What's funny about the passage, however, is the image of the self-important Mr. Ramsay clenching to squeeze out his next idea like a bowel movement. You don't need to know anything about science to know that you cannot sit down and force yourself to have a good idea. They come when you look at a problem from a new point of view. They come when you apply your experience and recognize a familiar pattern. Sometimes, they just come. The history of science—like the history of art, music, and mathematics—is full of stories of crucial insights striking from out of the blue.

Intuition is real. Cognitive psychology has measured it and defined its outlines in repeatable experiments. Some of this is obvious: everyone knows we can recognize patterns and perform simple tasks without thinking about them, and the definition of intuition is nonconscious thought. Consciously, we're able to think about only one thing at a time, but still we can drive and listen to a book on tape at the same time, or sing and play an instrument, or strategize and shoot a basketball. It turns out we can also think deep, complex thoughts without knowing it. There is no other explanation for an experiment in which a chess master defeated a weaker master in a five-second-a-move game while simultaneously adding up lists of numbers. Or for Japanese chicken sexers, those useful individuals who can judge the sex of a thousand newly hatched chicks in an hour with 99 percent accuracy; no one else can do it until six weeks after hatching. American apprentices learned the skill by watching the Japanese. But neither group could explain what they saw that told them a chick was male or female.

When my older daughter, Julia, was a tiny baby, I witnessed the moment she learned that she could affect the physical world. She was lying on her back under a bar with toys hanging down toward her, mildly amused by their movement, which was caused by the random flailing of her arms and legs. Then she caught sight of her own hand hitting a toy and the effect it had. She stopped and stared, then she took a deliberate swing and hit the toy again. Delighted, she began rhythmically swatting the toys with obvious intention. I watched each of my children test the density, strength, taste, and acoustic properties of every object they could get their hands on; every parent does, which is why our houses are so messy and our nerves so worn. I saw the surprise and experimentation of each of my children when they first tried throwing a ball. They started out without the slightest notion of where it would land. Later, they could pick up a rock from the side of the road and hit a stop sign with it, making a satisfying clang. The nonconscious thought required to execute such an operation is staggering. The brain has to recall the density of previous rocks and calculate the weight of this particular rock based on its observed size, calculate the distance and size of the sign based on a knowledge of perspective, then use all that information to calculate the angle and force of the throw to achieve a trajectory that will hit the sign. The expectation of a clang is based on knowledge of the properties of sheet metal. I suppose the brain even has to remember that it can affect the physical world, as I saw Julia discover.

I learned many things more complex than I could ever calculate with my conscious mind. I knew how big a nail I could pound into a piece of wood without splitting it. I knew how to trim sails and steer a boat over waves for maximum speed. I knew when to start gathering wood so the fire would be ready in time to cook the salmon before the kids got too hungry. When I was in my twenties and my grandfather was in his eighties, he beat me at a game of squash. As hard as I hit and as fast as I ran, he could still drop the ball an inch beyond my reach. Sometimes I could see how he had maneuvered me through a series of soft but perfect shots, but other times all I could do was pant. It's exciting to think how much you can learn by that age.

Cognitive psychology established as well that intuition is often wrong, sometimes insistently so. The phenomenon of attribution error that I mentioned in chapter 2 is an example: people believe that others are to blame for being victims, but when the same thing happens to them they feel innocent. Racism is intuitive, too. No one decides to be a racist through a logical thought process. Conscious rational thought avoids these errors because its chain of reasoning can be laid out in discrete steps and diagnosed. Put the steps in a scientific paper, and anyone can run through them. Mr. Ramsay's A-to-Z model of the world is difficult for science to escape. But intuition can be trusted when it can be checked. With enough trial and error feedback, nonconscious thinking can become reliable enough to trust with driving your car.

For Iñupiaq whaling captains, errors can be fatal, but the snap decisions they make for the safety of their crews are unlikely to be based on careful logical reasoning. The real world rarely allows sufficient time or information to apply rigorous analysis. We usually rely on judgment. A whaling captain has to size up complex, changing ice conditions in a matter of seconds. He needs to think faster and better about ice than any computer can, evaluating its strength, the position and number of the pressure ridge anchors, the direction of the current and wind, the shape of the bottom, the configuration of the cracks, and the patterns of movement that normally arise in similar circumstances. The lessons of experience allow a good whaler to recognize a situation rather than figuring it out.

People, and even many animals, are better at recognizing patterns in unique circumstances than computers are. The brain cannot process information as fast as a computer by a factor of millions, but its ability to apply patterns can bypass the need to process so much information. Advanced magnetic brain imaging shows how this works in chess grandmasters: amateur players analyze the game in the part of the brain for processing new information as they think it through, but grandmasters call upon long-term memory for patterns that match what they see on the board. A computer, IBM's Deep Blue, wasn't able to beat the world's best human chess player, Gary Kasparov, until programmers taught it to think more like a human and gave it libraries of human moves. Yet chess is, by its fundamentally finite and logical nature, a game played on a computer's home court, like John Henry taking on the steam hammer in a race of railroad building. The real world is an infinitely free, nonlinear system. It makes sense that people who simply watch the world know more about it than a computer could, and even that they know more than those who have analyzed it in a step-by-step fashion rather than experientially.

The Iñupiat developed a collective body of knowledge over a thousand years of subsisting from their environment. In a language perfectly suited to the problems it addressed, they held long talks in camp, in the qargi, and at the Volunteer Search and Rescue that synthesized what many people had seen over broad spans of time and space. New ideas and technology were adopted when they worked, and old ideas were discarded when it was clear they didn't help: for example, whaling crews abandoned their use of talismans and magical rituals when they saw that the Yankee shore whalers led by Charles Brower and George Leavitt succeeded without them. The insights that worked—the patterns that held true—were passed to the next generation. The structure of the whaling crew guided that process, with members gaining skill and responsibility as they aged. The elders, who knew the most, were responsible for the whole community. The Iñupiat learned at the start of the Little Ice Age that to whale they must share knowledge, strength, and food. One whaling captain's intuitive understanding of the ice was the product of many minds.

Jim Maslanik believed his mind was meant to work like an Eskimo's, to take in a lot of loosely related information and synthesize it into new ideas, but it was getting progressively harder to think that way. He didn't have enough time between all the proposals, reports, budgets, and meetings. He didn't get to talk enough either because, so short on time, he found himself relying on e-mail, even with his next-door neighbor. He would like to gobble books, but computers had made that harder, too. Jim used to wander through the library stacks, and when he was looking for a particular book he would flip through others on the same shelf and often find something he didn't know he was looking for. Now a computer zeroed in on the exact book he needed and a librarian retrieved it from space-saving shelves that collapsed together, eliminating aisles. "The first day I went there, I said, 'Oh, this is horrible.' There was a library assistant there who must have thought I was on drugs. Here was this old guy, ranting about this bookshelf thing." Once the library had been a sea of knowledge in which to swim; now one asked a question of Google and it gave back an answer.

Computers themselves had become too fast, complex, and full of information for a human mind truly to control, more like natural systems than machines. At IARC in Fairbanks, a model-comparison researcher named Uma Bhatt resorted to paper and intuition to help understand where GCMs were going wrong. She organized the psychedelic disks that the models produced—plots of temperature or other climate indicators on a world map—with glue and spiral-bound notebooks, gazing at them as if they were aids to meditation. Sometimes Uma would look at the mathematics in the models, too, at least the numbers that emerged from changes she made herself, but she knew she would quickly be lost wading into all the interacting calculations that flowed through the machine month after month.

"My adviser would keep telling me, work at building your intuition," Uma said. The goal was to be able to look at one of these plots and, like a Japanese chicken sexer, see something you couldn't see. Uma learned the skill well enough to correct even the adviser who had told her to develop it. "I looked at a plot and said, 'That's wrong.' She said, 'No, it's not.' Then she went back and checked, and there was a bug."

Unlike the Iñupiat, climate change scientists lacked the ability to share their intuitive insights. They lacked even a comprehensible common body of knowledge. The scientific literature exploded to the point that many specialists gave up trying to read everything published even in their own area. Libraries couldn't handle the flow—there were just too many journals. Virtually every journal became accessible through the Internet, and some existed only there, but the ease of searching from one's own computer was more than negated by the surfeit of articles a search on any general topic would bring up. The choice then would be to define the terms of the search precisely to catch articles on a narrow combination of concepts, or to trust the computer to put the most relevant articles at the top of the list. That meant giving to the computer an important component of the intuitive thought process, the job of seeing connections and patterns in new information. It also made it difficult to catch up broadly on a line of research. Some journals helped by publishing review articles that rounded up important findings on a particular topic in a few pages, with scores of references at the end. A review article could then stand like a navigational buoy in a sea of shifting information. But that couldn't solve the problem of communicating abstract concepts across disciplines. *Science* published an article that found seven inconsistent definitions of the important term *thermohaline circulation* in the scientific literature. The author concluded that in such cases, "What everyone thinks they understand may in fact be a muddle of mutual misunderstanding."

Norbert Untersteiner, the NARL veteran responsible for many key sea ice measurements, took delight in pointing out, with his Austrian accent and deep chuckle, how the competitive forces that drove scientists cast all commonly adrift. He plotted on a logarithmic scale the number of pages in the *Journal of Geophysical Research* from 1960 to 2000: they rose on a straight trend line, up by a factor of ten every forty years, which would put the journal at a hundred thousand pages by 2020. He also wrote a funny piece in *Physics Today* titled "Cite This Letter!" which pointed out, "We must remember that the primary purpose of publishing anything is for the author to be cited, and that the best way to get cited is to cite other people, no matter how trivial their work. It follows that there is little point in referencing anybody who will not reference you. With a few exceptions . . . this automatically excludes the dead and the retired, no matter what they did for us. The people who need citations are those of us who still need promotions and merit raises." Later, researchers published a real paper (as opposed to Norbert's joke) establishing that many papers cited in scientific literature were never read by the authors who cited them but merely copied from one list of references and pasted into another. Consequently, computer-calculated impact factors, the objective measure of a paper's worth based on how frequently it was cited, sometimes conferred fame on unimportant work that few had actually read.

Of course, Norbert himself was one of the most cited of Arctic researchers. His career in the Arctic dated back to the International Geophysical Year, when he came to the United States and had to fill out a lot of McCarthy-era paperwork. He had served as a draftee in World War II on the German side. The forms asked, "Have you ever been a member of an organization that advocated the overthrow of the United States government?" Norbert shrugged: "I had to say yes." Now in his mid seventies,

Norbert had the role of scientific elder, guiding younger scientists with his store of memories and understanding of how science works. During a visit to IARC, he was available simply to talk—a delightful experience for anyone who could tear away from the computer for long enough—helping researchers refine their ideas, keep their work in context, and avoid repeating what had already been done.

Avoiding repetition was not as easy as it would seem. A scientist planning even a narrow project might need a year of free time to conduct a thorough literature search to find out if the research was new. That was not practical, so a lot of work did get repeated, according to Bernie Zak, the atmospheric physicist. The flaw might go undiscovered until the very end, at the peer review phase of publication. Bernie told me this in Barrow after hearing a talk on a project that was still in the field; I'd been out with the scientists who were doing this work in –17 degree F weather and a strong wind, struggling with balky equipment that they didn't really understand. One of the NARL veterans leaned over to Bernie during the talk and said the same topic had been studied twenty years earlier.

The highest rewards in the scientific culture go to new, individual discovery. I met researchers in Barrow who looked on the unmarked white snow as their chance, as if they were setting out alone into the uncharted unknown. But decades of research had made it harder to discover something new. If you counted the Iñupiat's knowledge, it was harder yet. This team had used a satellite to identify tundra features and had even given some of them fanciful names, then set out to categorize and study the areas. I met them preparing with trepidation to go to a distant site in cold, stormy weather. Yet the Iñupiat already knew some key characteristics of these areas and had given some of them names long ago; the object of that day's expedition was the far point of occasional Eskimo snowmachine races.

But tapping the knowledge of elders, either Eskimo or scientific, wasn't always easy for a newcomer. Jim Maslanik said most researchers would not simply pick up the phone and call a leader in their field and ask a question. The process of proposals and grants that paid for most research promoted competitiveness. As soon as BASC got word that Ted Stevens would fund a new building, scientists working in Barrow began worrying privately about whether the facility or its maintenance would siphon money away from the programs they applied to. In the chase for grant funding, another scientist's gain might be your loss. In private little conversations I heard cutting criticisms of projects; it was sometimes hard to tell valid concerns from envy. Scientists did enjoy mutually sustaining communities that shared knowledge among people who had worked together and formed a personal bond. But these groups were small.

Competitiveness drives science forward, and science has come a long way since the United States adopted the proposal system after World War II. But competitiveness has also left important things by the wayside. Science failed to make the long-term observations that would greatly simplify the search for a climate change fingerprint today. Much of what we have—other than automated measurements from the weather service and such—came from farsighted researchers who accumulated their own long-term data sets by doing many individual projects in the same place with the same processes. Oceanographer Jackie Grebmeier and permafrost researcher Tom Osterkamp were prime examples. George Divoky, a bird researcher who worked on a barren barrier island east of Point Barrow, returned for twenty-five seasons to the same place even though he often had no funding and lived in an unheated tent. But heroism and foresight are rare, so the system got what it paid for: a glut of papers on short-term projects and a lack of coherent community knowledge about the whole Arctic system.

NSF administrators, who are scientists, understood the problem and tried to push research in the right direction by setting up initiatives—umbrella programs to spur research on an overriding theme. The NSF's Arctic System Science Program was such an umbrella, dedicated to understanding the Arctic system and climate change, with other umbrellas under it: for interactions between ecology, water, atmosphere, and so on; for human activities, the environment, and resources; and for climate change. And under each of those, nests of other programs and projects with subcomponents. It took a lot of meetings to set up all these programs within programs, to agree on goals and how disciplines should work together. Various parts of the NSF had to work together as well as the research community, so many committees and conferences were convened. Acronyms bred like germs until a newsletter intended to help make it all clear said things like: "These investigations have contributed both formally and informally to several international programs, such as BIOME 6000 (the

Global Paleovegetation Project, an IGBP effort) and its Arctic component PAIN (a joint European-Russian-United States research project co-funded by Paleoenvironmental Arctic Sciences [PARCS]), and CircumArctic Paleo-Environments (CAPE); the circumarctic element of IGBP-PAGES."

Scientists proposing projects for funding through these interlocking programs needed to understand how to make their ideas fit into the framework, which sometimes meant developing projects to advance program goals and sometimes meant pitching what they planned to do anyway in terms that were currently hot. Figuring out all that organization, attending the meetings, writing and reading the plans, tailoring the proposals, keeping track of HARC, PARCS, ARC, ARCUS, ARCSS, and ARCMIP—it all took up a lot of brain space. "It's amazing how people who are paid, supposedly, to think, get so little time to do that," Jim Maslanik said. "People say, 'I take my shower in the morning and that's my fifteen minutes for thinking.' I can't do science if I can't just think."

Scientists themselves push these programs, however, because programs are the best way to cooperate and compete at the same time. Jamie Morison of the University of Washington took on the task of organizing the broadest of all the Arctic umbrella programs, called SEARCH, for Study of Environmental Arctic Change (it grew out of ARCSS as an element of CLIVAR . . . oh, never mind). The concept was to find out why Arctic climate warmed so drastically starting in the early 1990s by coordinating all the Arctic science resources of the federal government. Jamie said that long-term monitoring had actually declined since the changes began. The navy mothballed submarines of the so-called Sturgeon class capable of operating and taking measurements under the ice. Scientists had been allowed to ride along on navy cruises, and one of the findings from the project had been that the Arctic sea ice had thinned by 42 percent. But when the navy didn't need the subs anymore and offered to give one to the NSF with eight years of nuclear fuel left on board, the operating costs were more than science could afford. The Russians quit their extensive Arctic observing systems, too, and the big ice camps of NARL days were no longer being mounted. Jamie Morison's science committee put together a plan with big, ambitious hypotheses to test, a component for long-term monitoring, and a budget of $120 million a year, almost double everything the NSF was spending on the Arctic annually. Jamie realized his group was probably dreaming too big, but it was important that the money be new. The ten federal agencies that were involved demanded new money as part of the arduous process of getting all involved.

The first step to selling the idea would be to give it a name that would catch on. "We keep calling it 'the complex of interrelated Arctic changes that have occurred in the last ten years,' " Jamie explained. "We needed something shorter." He called a meeting and invited Caleb Pungowiyi, an elder from Kotzebue who had been Henry Huntington's mentor in studying traditional knowledge. Scientists loved Caleb because he came across with immense dignity—calm, intense, and erect—but he also knew how to speak their language, melding scientific jargon with poetic terms from the Siberian Yup'ik culture of his childhood. Jamie asked Caleb for the Eskimo word for "change," but Caleb said it was too long, without saying what it was. The committee process then came up with Nuevo Omani, which they interpreted as a bilingual "new tomorrow." Jamie dropped the Nuevo, but Omani still wasn't quite right linguistically, and besides it sounded too much like Bon Ami soap. Then Jamie heard of a chain of Japanese restaurants called Omani. Finally, the committee settled on Unaami, Yup'ik for "tomorrow." After a couple of years of heavy use by SEARCH members, however, I never heard anyone else utter the term.

Getting the money was harder than picking the name. The top people on Arctic issues in the bureaucracy were middle-level people within their agencies. For them, the agencies' agreement to cooperate was a major accomplishment in itself. Obtaining money from Congress and the White House wasn't their job. But Jamie Morison and his academic colleagues weren't much better equipped. As the Bush administration prepared its budget in early 2002, he was living on rumors. "That's where we are now," he said. "We're in the walking-the-halls-of-Congress stage. Nobody knows how to do that. The agencies can't do it." As the day for release of the budget neared, members of the U.S. Arctic Research Commission went to bat for SEARCH. One was Jack Roderick, a former Anchorage mayor and Ted Stevens's old law partner. He called Ted and asked him to put in a request with the administration not to cut the agencies. For whatever reason, the money did come through. Commission member Mead Treadwell, another Anchorage friend of Ted's (Mead's mentor, former Governor Wally Hickel, appointed Stevens to the Senate in 1968), set up a press conference for Jamie in Anchorage to announce funding for the program, \$2 million a year for NOAA and \$5 million a year for NSF for five years. Well short of the dream, but enough to get started.

I think I witnessed the year when Barrow's Iñupiaq leaders accepted that climate change was real and one-way. Most of the elders already believed, but the executives I met at the Hotel Captain Cook still had thought it might just be a cycle. Over that year, however, many people accepted that, even if the warming was the peak of a cycle, it was a higher peak than had ever been seen before. In the midst of the wet thaw of the terrible spring whaling season of 2002, Oliver Leavitt and Richard Glenn believed. They told Ted Stevens; he told me he believed it because his Eskimo friends said so. Richard conceded that some of the evidence was irrefutable. He stood out on the ice and said, "The climate is changing. The ground tells us it's warmer than it used to be, and so do our old people."

Richard and Arlene's new baby was born in the early summer, after the whaling was over, at a hospital in Anchorage. I took my three older children to visit and brought along my whaling photos. It was strange to see Richard and Arlene alone with tiny Joanne in that little room. To me, they should always be flowing effortlessly through a big group of people. The outcome, however, was good: Joanne became a delightful baby with that same exaggerated grin that Richard liked to pull off, a grin so wide and sharp that it closed the eyes.

The next fall, when Joanne was a good-sized baby, I met Richard and Arlene for a taped interview in one of the dusty labs at the ARF. Arlene carried Joanne in her parka hood, then sat and nursed her under the colorful fabric of another of her beautiful handmade parkas.

Richard: "Your question was what has happened since you started talking to me, to today, that convinced me it was getting warmer? I don't know. I'm trying to think of what position changed. I was always looking forward to a back side of this cycle. I'm still hoping that that back side is coming. And if it doesn't, then we have to change. If it's part of a one-way progression, then life is really going to change. But even the experts are out on this. The people who are closest to the study still say they just don't know. If the timing of your book was two years off in one direction or the other, you might have seen a totally opposite spring and fall season than we've seen now. So I'm nervous about attributing some overarching trend to a relatively small time period."

Charles (me): "Yes, but I'm not writing a scientific book. There will be science in the book, and hopefully I am going to get all that stuff right. I am certainly going to try hard to. But the story itself is full of metaphor."

Richard: "Fit the metaphor. The last sentence of your book can talk about the read-write CD-ROM that my daughter is using to download music for our weekend get-togethers. She's even writing, onto a CD-ROM, Eskimo dances from way back when. But you can't do that. Don't metaphor me."

Charles: "Too late."

Richard: "If it sounds good, you'll use it."

Charles: "You talk about, 'If this change is a one-way thing, that life is going to change a lot.' "

Richard: "Big change."

Charles: "What do you mean? What kinds of things do you think about that would change?"

Richard: "If we start losing the spring season we have to totally rethink ice safety. The rules change. The rules change season to season anyway, but this would be a bigger shift. Things that were true for fathers won't be true for sons, and so it will always be experiencing something new."

Charles: "That's kind of been the case for the last 150 years anyway."

Richard: "Oh yeah, the culture has changed, always. But there's always been some things. The ice on the lake will get five to six feet thick every year. Or ice that's accreted to the shoreline with enough pressure ridges is probably going to stick around. Those kind of things, those little rules of thumb are going to change. And that will change how you travel, how you hunt, how you stay alive."

CHAPTER EIGHT

■

The Camps

I DROVE TO THE ARCTIC in the constant brightness of midsummer. Early one July morning, I pulled my SUV out of the driveway in Anchorage with thirty gallons of extra gas and a spare wheel tied on top, drove north over the tundra slopes and through the craggy rock of the Alaska Range and Denali National Park, stopped to say hi to friends and eat dinner in Fairbanks, then drove on through the stunted black spruce forests of the permafrost river valleys, across the Yukon River, and up onto an endless mountain plateau of tundra and granite outcroppings like statues, where I pulled over and slept. In the morning, all remained bright and huge, still no other cars or people, the highway leading irresistibly on, and it seemed I had never done anything else but pull this infinite string, this road, that unwrapped this infinite landscape. Presently, perhaps once or twice in an hour, a distant rooster tail of dust rose on the horizon and slowly neared. Fancy that: other people did exist.

The Dalton Highway was built in the 1970s for the construction of the trans-Alaska pipeline and it remained a true wilderness road, without development over roughly five hundred miles from Fairbanks to Prudhoe Bay except a truck stop at the Yukon River and another at Coldfoot. The sinuous line of the dusty road and the silvery thread of the pipeline—it was four feet thick and stood up on legs—seemed to tie the country together, a subject for the vistas. Away from the human mark this wild country can be hard to absorb, hard to remember: a day hiking off trail or flying a small plane develops like a child's story or a retold dream, full of events but without a plot or a conclusion. There is no context but oneself in the constant repetition and constant uniqueness, which, like the paisley patterns of fractal geometry, exist on every scale, from the forest of tundra heather three inches high to the spires of the Brooks Range. The highway and the pipeline added a unifying theme. The landscape's story gained a beginning, middle, and end. Photographs of this region usually show the pipeline and road. It's harder to take pictures without a person, animal, or man-made thing in the frame; it's hard to decide where to point the camera because the viewfinder always leaves out most of the beauty.

At Atigun Pass the road climbed over the top of the Brooks Range, where, at 4,700 feet elevation and several degrees north of the Arctic Circle, nothing grew on the vastness of ice-carved rock. On the far side, the North Slope began. The foothills came first, as round and long as hips, a gathering of voluptuous giants lying on their sides, barely clothed by tundra. The road draped as lazily as an untied sash over the body of the earth. At any point the car stopped it would have been possible to hike over dry tundra ridges, totally free. Then there was a branch in the road and, in one of the rounded dips, a lake, and by the lake a scattering of rectangles. Buildings on a gravel pad. The Toolik Lake Field Camp.

Inside one of the trim, modern modular buildings, several white plastic folding tables had been pushed together into one, nearly filling a lab, and around that big table sat a dozen young adults and one middle-aged man. On the table, hands were pulling apart clumps of dirt and tundra the size of volleyballs, picking out each leaf, stem, root—all these tiny parts—and sorting them one by one into piles and metal trays. The workers had just enough room for their own piles, elbows pulled in to avoid their neighbors. This was called "the pluck." It looked like a scientific sweatshop, but the work was skilled. These were the most promising of ecology graduate students and postdocs, able to be here only because they were studying at the top of their field. The pluck required them to recognize each species type in the miniature tangle of tundra growth, to separate the stems from the roots and the roots from the runners, to separate alive and dead, and sometimes to identify a plant two inches tall by its roots or bare stem alone. This work required intense concentration, and on this particular day the work had started at 8:30 a.m. and would go past 11:00 p.m. It would continue with roughly those hours for more than a week. I found a spot to perch and ask questions, overcoming an urge to run. After absorbing 750 miles of Alaska, watching mountains blur together like blades of grass, my mind balked at switching to the scale of examining individual leaves, as if I were an orbiting astronaut suddenly asked to count grains of sand on a beach.

Gus Shaver presided at the table. His physical build suggested density and power, the sort of body it would seem impossible to pin behind a card table with a pile of tiny leaves, yet his presence kept the group together. He was a senior scientist at the Ecosystems Center at the Woods Hole Marine Biology Lab (known as MBL), where most of the students and researchers currently in camp were based. Many of their careers still followed like ducklings in the wake of his grants and projects. Gus had come north with the ecology pioneers in the International Biological Program; he had shared a lab in Barrow with Walt Oechel in 1972. But his approach and that of most of the other scientists at Toolik diverged from Walt's work in Barrow. Walt's towers and low-flying airplane sniffed the carbon dioxide in the atmosphere above the tundra; Gus and his colleagues were seeking the paths of carbon and nutrients by studying plants and earth directly. "What he measures is the sum of hundreds and hundreds of processes. What we're doing is trying to understand some of the processes that go into that," Gus said.

Ecology seems to me the hardest science. Ecologists are chemists, physicists, biologists, and mathematicians. They study complex living systems with many variables to learn the flow and combination of some fifteen chemical elements and how external changes alter those interleaved paths. For example, Gus, with Terry Chapin of the University of Alaska Fairbanks and other colleagues, found that, while increased CO_2 in the atmosphere did augment tundra growth, the boost was only temporary: soon the plants ran short of mineralized nitrogen (the nitrogen released by soil decomposition). A finding like that can take years of complex experiments to tease out of the system. The plots harvested for the pluck when I visited had been set up thirteen years earlier. After everything was sorted and dried, each part of each type of plant would be measured to find its composition of carbon and nutrients, and that data would feed into models showing how different treatments of the plots—warming, shade, fertilizers, and so on—affected the processes. Many factors changed at once.

Warming could affect the growing season, protective snow cover, moisture, soil temperature, activity of soil microbes and fungi, the mix of plant species, how they used nutrients, how they shaded one another, and so on. The papers that came out at the end of the work were technical and not easy for a nonspecialist to understand. Broadening those findings to the landscape or global scale was harder yet. And what if you made a mistake back at the start? There were various aspects of the study design that Gus wished had been decided differently two decades earlier, but it was too late to change now.

As tough as ecology is, it's not a way to get rich or gain status. Gus estimated there were only about a hundred experimental plant ecologists in the United States, and at least one of those I met (not at Toolik) didn't even want to be called one: he thought the image of ecology was too fuzzy and feminine and preferred to be called a biogeochemist. At the pluck, all the ecologists besides Gus, an energetic German postdoc, and a quiet young man specializing in butterflies, were young women, and most fit the stereotype pretty well: vegetarians and granola eaters who wore natural fabrics and were healthy and fit, outspoken and unaffected. The group talked about people and minor controversies in the camp as they worked, occasionally helping one another to identify particularly tough samples or asking Gus. They had their own projects relying on data harvested from the plots. Gus was a mentor and manager, plucking away—he didn't expect anyone to work harder than he did—and answering questions with unquestionable authority. Someone came back from the dining hall with a big mixing bowl full of ice cream laced with chocolates and other treats: a Vermonster, it was called, and the rule was that it would be passed around the table with everyone taking bites in turn until it was gone. Gus took his share. He handled the room's generational and gender differences graciously, trying not to complain about the techno music on the boom box but smiling gratefully when they put on Van Morrison for his benefit. Back in Falmouth, his wife and fifteen- and seventeen-year-old boys were involved in a sailboat race that day.

Formally, Gus was coordinator for the dry land, or terrestrial, studies of Toolik's Long-Term Ecological Reserve, one of an international network of research sites. Toolik itself, however, was owned and operated not by the Woods Hole MBL but by the Institute of Arctic Biology at the University of Alaska Fairbanks, as I was urged to mention prominently in anything I would write. Scientists explained quietly that there were long-standing jealousies between the operators from Alaska and the users from Massachusetts. Arranging an invitation to the camp wasn't easy. I broke in with the advice of a science elder I had met in Barrow, who had helped found the camp back in 1975.

But once I was in, they treated me royally. Few remote outposts in Alaska, including pricey wilderness lodges, were more comfortable. Toolik was like Prudhoe Bay's oil facilities in the way it brought soft civilization to the Arctic. Facilities were new, well built, and expensive. More than a hundred researchers worked and lived in this glorious lakeside setting, their every need provided for—plentiful and expertly prepared meals, snacks always available for the taking, shiny, well-equipped labs, brand-new showers, a large sauna on the shore, and electricity from diesel generators operating all kinds of equipment, including computers and telephones connected to the outside world through a fiberoptic Internet link. The camp manager had received a bunch of Iridium satellite phones and, not knowing what to do with so many, handed them out to anyone who was going for a hike. The NSF picked up the tab—researchers paid only the incremental cost of their visits—and running Toolik cost a lot. For example, waste water could not be disposed of at the camp because it might pollute the lake that the scientists were studying, so all the sewage and all the gray water—the effluent of every shower, dish tub, or mop bucket—was trucked 367 miles back to Fairbanks.

Toolik hadn't always been so comfy. It started as a trailer surrounded by sleeping tents. It grew slowly for years. A Toolik culture developed, with Sunday hikes and outrageous Fourth of July parades. With the need to conserve water, everyone in camp would get clean by taking saunas three times a week, holding nude impromptu seminars on their work (the showers showed up only in 2002). The camp became an insular world bounded by the edge of its gravel pad on the tundra, linked to the outside only by two unreliable phones and a Sunday *New York Times* that came on Wednesday and was kept wrapped up until the following Sunday. Summerlong residents returned home having missed the collapse of the Soviet Union or wondering why everyone was talking about O. J. Simpson. The population of the camp was older then because of the way projects were billed: they paid the full costs per person, per day, so PIs came themselves and brought productive postdocs and technicians, not students, to get their work done as fast as possible. The atmosphere of a world away from the world—with the endless daylight and tedious work—broke the spell of social inhibitions. Many relationships sparked to life, some of which stuck and turned into marriages. When I visited, people were talking about embarrassing party photos that had shown up on the Internet of a Toolik veteran who had met his wife there. An older scientist shook his head and said, "Some things should stay on the pad."

With the NSF's big investments, many more students, even undergraduates, filled the sleeping trailers and camping area. Older Toolik veterans pined for the good old days, when everyone took saunas together, meals were small enough for group discussions, and the camp culture was more serious. Now, the Internet had erased the isolation, meals were big dining hall affairs, and men and women segregated more at sauna times—some were even crass enough to use the new showers instead of the sauna. When I visited, the camp felt like a college dorm transported to the Arctic, with partying and sex a major topic of conversation. The situation might seem ideal for a single male grad student outnumbered by sophisticated, attractive women with little to do in the evenings, but there was a hazard to living in such a small community. According to a popular story in camp, a newly minted couple had been seen going into a room together, and then she was heard through the thin walls, saying with consternation, "Is that it?" The words became a catch phrase around the camp: "Is that it?" But some older scientists were not amused. Matthew Sturm had spent a night at Toolik just before I arrived. He said he never saw his roommate, despite going to bed at midnight, but a nineteen-year-old who had been drinking all night showed up in the morning. "I've spent a little time at Toolik, and it's not my cup of tea," Matthew said. "I'm too old for Toolik. I've spent some time there in the dead of winter, when it's deserted, and I've enjoyed it." But in the winter, the camp was closed.

Matthew, Ken Tape, and their colleagues were working on their shrub photographs. Matthew, an expert canoeist, led a weeklong expedition on the Chandler River, which flows from the Brooks Range north across the North Slope, to inspect the shrubs shown in the aerial photographs and improve the team's ability to identify the species in the pictures. "There are times the shrubs are just towering over your head. It's not the picture you have of the North Slope," Matthew said. He was a controversial interloper in ecology with his theory that snow aided shrub growth. His idea was that shrubs and snow drifts related in a positive feedback loop, with shrubs gathering snow that then kept the ground warm and further aided shrub growth. As a snow physicist, he admitted he kept quiet when Toolik ecologists talked to him, pretending he could understand. He had a member of their community on the shrub project, and Toolik cofounder Terry Chapin was a coauthor of Matthew's key shrub paper in the *Journal of Climate*, but still the idea had run into serious attack, and it took two years to get the paper published. Matthew's style of capturing patterns of change on a broad scale appeared to challenge the ecologists' tradition of process studies that measured changes in every leaf and twig. "I don't think the way I function is to sit at a plot and nurse it for twenty years," he said. "That's not how I work. That's pretty tough stuff. Plant ecology is profoundly confusing, because there are so many things interacting."

At the pluck, when I asked Gus about Matthew's shrub theory, the conversation cooled. He chose his words even more carefully—in fact, he seemed suspicious of me for the rest of our time together. Gus and Matthew both made a point of saying how they admired each other's work, but their disagreement was important. Gus and his colleagues had shown that warming plots with greenhouses exploded shrub growth, especially dwarf Arctic birch, *Betula nana*, a tough, scratchy bush with small, thin leaves. Greenhouses of plastic sheeting, removed in the winter, became a solid tangle of three-foot-tall birch in a decade. Underneath, the moss and shorter, thick-leafed tundra plants lost out to the bushes. Gus believed that summer warmth explained the advance of shrubs, and he did not buy the idea of a positive feedback loop with snow drifts in shrubs leading to more growth. The shade cast by birch leaves should cool the ground in summer, a negative feedback that would overwhelm the positive winter effect. Usable soil, the active layer atop the permafrost, did thaw deeper under the bushes, but that might be the reason the shrubs were present, not a result of their ability to accumulate snow. It was a debate about which was more important, winter or summer, and many scientists in camp had adopted their own opinions. Gus, who had been studying the Arctic for thirty summers without ever visiting in the winter, didn't believe the winter was as important.

"Gus studies growing plants and Matthew is a snow physicist—what would you expect them to say?" said Josh Schimel, a soil scientist from the University of California, Santa Barbara, and a Toolik veteran, who was working with test tubes of dirt in one of the labs. Not enough was known to do more than form hypotheses, he said. "I don't understand tundra soils, and there's nobody who understands them better than me."

What was happening in the soil held the key. That was where the biological activity occurred that controlled growth on the surface. A lack of nutrients limited plant growth. To make nutrients such as nitrogen into forms plants could use required underground decay, or soil respiration, by microbes. Warming accelerated soil respiration, releasing more nitrogen for plants and speeding growth. Birch happened to be especially efficient at incorporating extra nitrogen into new wood, so it won the growth competition over other tundra plants, shaded them, and took over the land. What made Matthew's work exciting, Josh said, was that it showed that snowdrifts on shrubs could greatly lengthen the time when the ground was warm enough for soil respiration: fifty days longer if the limit was –6 degrees C, or even longer if, as Josh believed, respiration could go on down to –10 degrees C. Arctic soil could be decaying under the shrubs most of the winter. But the question remained how much soil respiration could happen at those low temperatures. Josh was collaborating with Matthew using a nitrogen isotope as a tracer to measure the importance of increased snow cover on nitrogen availability.

Josh wanted to think broadly, to teach his mind to take winter into account as well as summer, escaping from the limits set by his discipline and the academic calendar. It took time in the field to build intuition about how a system worked, and since fall was the end of the academic field season, scientists had come to think of it as the end of the story. Language also kept scientists segregated from new ideas. "Language does control how you think, and if you don't have a word for it, you won't think of it." Josh liked coming to Toolik because so many scientists were thrown together there to collaborate—a result of the NSF's Arctic System Science Program.

"The challenge of doing science on these systems is scale. You need to scale down to understand, but then you need to scale up to make it fit together. We have to keep from falling into these reductionist funnels," Josh said. "To work at the larger scales you need to be able to take your glasses off and say, 'It looks OK to me,' once you can't see all the details."

Terry Chapin believed the winter-summer debate would end when the research was synthesized in a way that would allow everyone to see the system together. He was in the winter soil warming camp: because winter was so long, more than half the nutrient breakdown happened then, thanks both to the microbes' work and to freezing, which broke apart root and microbe cells, releasing their hoard of nutrients. Summer shade might cool the ground, but the leaves falling from shrubs would decompose more quickly than slower-growing tundra plants, and that would offset the shade effect.

Terry was one of the most respected scientists in the field for his ability to bring together many strands of research into a whole picture, and he had taken the shrub issue to the broadest scale. He predicted increased shrubs and forest would contribute to climate warming by reflecting less of the sun's energy than the tundra they replaced. The albedo, or reflectiveness, of different kinds of plant communities depended on the details of how they grew: dead sedge leaves in wet tundra made it the most reflective ecological type while the dark colors and complex canopy of forest tundra made it absorb even more energy than shrubby areas. Generally, taller plants absorbed more energy because light bounced around among their branches; in comparison, a plain of heathery tundra was a flat mirror. An unfortunate coincidence: the progression of vegetation due to warming would create a landscape that absorbed more energy from the sun, further advancing warming. A modeling project led by Terry Chapin with Amanda Lynch, the University of Colorado atmospheric scientist, found that extending the vegetation changes that already had started in the Arctic would lead to regional warming equal to a doubling of atmospheric CO_2. That warming would probably further accelerate carbon dioxide release from formerly cold and frozen ground. Everyone seemed to think CO_2 release would increase, but no one wanted to guess how much or for how long.

Gus Shaver and Matthew Sturm didn't want to talk about those matters. Matthew said, "The real big-picture guys, they never even go out into the field. It would just confuse them." Or discourage them. Picking leaves off a single piece of heather might be too hard if you thought about how each tiny step related to understanding global change as a whole. The pluckers didn't even want to think about how much work they could do in a day. To survive such a job, you had to assume it would last practically forever, concentrating only on the next twig; the moment you started checking the clock, time would stretch horribly. I found it nauseating to think of all this work, all this intellect, feeding someday into a GCM as lines of equations.

Dinnertime brought a respite, a highlight of the day when the whole camp gathered like a community, loaded paper plates from a buffet, sat together at long tables, and complained about the food. The camp cook was doing her best, but this was a tough and unfamiliar audience. Normally Alaska's remote work camps house workers in the construction, oil, timber, or mining industries, mostly men tired and hungry from long hours of physical work and ready for grilled steak, fried potatoes, and seafood, rich side dishes with plenty of cheese, cream pies, and the like. Toolik's tastes had as much in common with that menu as a slender, highly educated Woods Hole leaf plucker had in common with a burly itinerant lumberjack. The food was high cuisine in the camp-cook culinary vernacular, but the vegetarians, perhaps a third of the population, were unhappy. One Sunday dinner during my visit included sautéed scallops, roast duck, baked Brie, and charcoal-grilled asparagus, done to a turn, plus a dubious vegetarian casserole. Camp managers had gone to the expense of hiring a special vegetarian chef, but that spot was currently vacant and the regular cook was trying to please everyone. Over dinner a young vegetarian woman told her friends the amusing story of volunteering to "help" the cook rather than only complaining, a didactic mission that swiftly had gone awry. The cook, a woman more than old enough to be her mother, had welcomed her help enthusiastically and put her to work right away preparing something called vegetarian taco lasagna. The name of the dish alone was enough to produce a round of wicked laughter at the table. The story was told well: the idealistic plan of opening the cook's eyes, the reality of being trapped loading grated cheese into a huge baking pan. The concoction that emerged wasn't positively bad, but there was plenty left over. The vegetarians I sat with that evening didn't eat much before their plates were consigned to the incinerator and they returned to work in the labs.

From May 24 to July 20, the sun never set at Toolik Lake. It was hard to stop working, and once work was over it was hard to go to bed. Alaska summer has been compared to a drug, like coffee or amphetamines, but no drug was ever as powerful and long-lasting as an Alaskan summer, especially for the young. Sleep became a stupid irrelevancy. Living things vibrated and the fresh odor of headlong, twenty-four-hour growth filled the air. Late at night the air cooled and calmed and infused itself with the flavors of sap, dust, and dew and the breath of a smooth, misty lake. The sun had its eyes half closed, peering over the northern horizon, delivering across the pole a strange, timeless light that made objects dimly glow. When you are young amid this strange light and this exquisite air, you must do everything; there is no other place or time. I remembered that smell; sometimes I would catch a whiff even now. After midnight in my bed at Toolik, I was not surprised to hear a party commencing on the pad outside, under the pastel northern sun, and I felt nostalgia and envy.

In the morning, an aquatic scientist I had planned to interview asked for a postponement. Someone had spilled tequila all over her lab at a party that went on until 5:00 a.m. and it was too bad a mess for visitors. A sign showed up on the door of the dining hall requesting quiet around the sleeping quarters after midnight. At lunchtime I happened to sit with a Woods Hole research assistant named Dirk, a skinny, close-cropped Australian who wore wraparound dark sunglasses and seemed to be having a grand summer as a self-appointed director of recreational activities. He was excited about the previous night's discovery: chariot racing. The camp had a couple of plywood luggage carts that could be attached to the back of mountain bikes. The buildings on the pad made a fine oval course for Ben Hur–style races. Dirk was planning a championship race for that night, with rules and a single-elimination tournament bracket.

Diane Sanzone, a postdoc from Woods Hole, was in camp for a brief break from her summer on a remote mountain stream in ANWR, where she was experimenting to see how stream ecology responded to a nitrogen isotope fertilizer. It was a remote and beautiful spot, and she was eager to show pictures, her eyes bright with the disorientation of wilderness decompression. Diane had learned to use a shotgun that summer for bear protection. On July 5, a snowstorm had crushed her team's tents. The experiment was working beautifully, turning a cold, biologically poor Arctic system into a mossy stream that looked more like it came from the temperate forest. Toolik had carried on such stream fertilization experiments for many years; on one waterway, paid fly fishermen angled for Arctic grayling every day to advance a fish-tagging experiment. (Their complaints of slow fishing proved people are always capable of complaining about their work.) Fertilization could simulate the effect of warming; as permafrost melted it could inject big blasts of nutrients into streams. Perhaps Diane's experiment could also simulate disturbances from ANWR oil development.

The typical view of Toolik people on the ANWR oil issue was, as one would expect, not in sympathy with the prodevelopment view of the George W. Bush administration, the Alaska political establishment, and the majority of Alaska residents. Ted Stevens was so sure that North Slope oil development had been beneficial to the state and harmless to the environment that he commissioned an impartial report by the National Academy of Sciences to establish its cumulative effects. He was angry when the report cited harm to wilderness aesthetics, caribou, and the Iñupiat. The authors discounted the large increase in caribou around the existing oil fields by saying it might have been even greater if not for oil development. Some points in the report seemed biased against ANWR development and less than serious scientifically, such as a suspiciously New Agey statement by an anonymous "Arctic elder" inserted to represent the views of the Iñupiat, whose general sentiment of support for drilling was quite different. Other criticisms were undeniable: the oil field seismic trails, gravel roads, and scattered metal buildings did not make the area prettier.

But Richard Glenn pointed out in prepublication comments to the report's authors that even the aesthetic impact of past oil development depended on your perspective. He wrote:

> The statements about visual effects are written from the point of view of a person who is amidst the infrastructure, and who has arrived there perhaps via airplane, bus and road or some such means. It is the feeling of one who accesses the oil field infrastructure while looking for wilderness. If the point of view was switched to someone coming in from the tundra, say, by snow machine, then the stigma effect of the infrastructure would be different. You can cross a lot of tundra traveling south or west of Prudhoe Bay before you realize the presence of the infrastructure (and there admittedly is a lot of it). If you do not believe this then try it.

The report detailed research showing that caribou avoid roads, but Richard pointed out that during times of insect harassment—major stress times for caribou—gravel roads and pads were their favored habitat as a relative haven from bugs. The final report mentioned but didn't explore that point. The idea of gravel pads enhancing caribou habitat might have sounded sacrilegious, but the idea of enhancing wildlife habitat by making ponds was well accepted all over the United States. Of course, ponds look better to people than gravel lots.

We are conditioned by the national parks to link beauty and environmental value, a prejudice that makes as much sense as thinking attractive people form better friendships than plain people. The land that became parks in the Lower 48 included a smaller proportion of productive wildlife habitat than the land that instead became flat, boring farms and ranches. Those parks, relatively tiny natural reserves in fractured ecosystems, owed their survival to the features that people valued aesthetically—mountains, canyons, caves, geysers, and so on. At Toolik I met a graduate student who was openly excited about seeing ANWR for the first time, the place she had heard so much about. I asked which part she was visiting: ANWR is larger than ten states, ranking in size between South Carolina and West Virginia, and includes a swath of the Brooks Range, the foothills, and the flat coastal plain where oil exploration was proposed. When she returned, she smiled sheepishly, disappointed. She said ANWR was just more wet tundra and mosquitoes, the same swampy terrain that stretched for hundreds of miles east, west, and north of us. Even as an environmental scientist, she expected special nature to be more scenic. But ANWR was no nature museum; it was the real thing.

Gus Shaver and Heather Rueth, a Woods Hole postdoc, set out after breakfast in rubber boots along the narrow boardwalk planks that led across the lake's inlet stream and up the far hill, among small greenhouses, snow fences, weather instruments, and Plexiglas cones, following branches that split here and there to arrive at the edge of various experiments or to continue over the top of the hill. Although the ground was mostly dry, the boards helped keep feet from trampling years of work that looked like just more tundra. Gus and Heather stopped at a certain set of plots to harvest tundra tussocks for the pluck. When the experiment began thirteen years earlier, the design had called for harvesting plants within wire squares called quadrats that measured twenty centimeters (eight inches) on a side, but the shrubs had grown in such profusion that it was difficult to figure out which branches belonged in an imaginary twenty-centimeter column three feet off the ground.

Gus said, "Man, this is hard, Heather."

"I'm glad you agree."

"We're stuck with this quadrat size because that's what we started with," he said.

Similarly, the experiment started with a summer greenhouse of plastic sheeting, which warmed the air inside drastically and also cut off the wind, increased the humidity, and lessened the light. Newer experiments were using heating coils buried in the tundra to test warming with fewer complications, but Gus said the data from the greenhouse plots was still valuable, because other manipulations and modeling could pull apart the various effects.

Heather's own project, funded by an NSF grant that Gus wrote, sought to clarify the summer and winter debate under the birch shrubs. The study's hypothesis was that summer shading produced a negative feedback that cooled the ground, and that leaf litter wasn't sufficient to overcome the reduction in decomposition caused by the shade. Heather planned to overwinter in the Arctic, taking soil temperatures and measuring the availability of nitrogen through the year. When I asked if this meant she favored the summer side of the debate, Heather lowered her voice so Gus wouldn't hear. "I don't really know. I come from Colorado, and I think winter might be really important."

The brilliant blue sky, warm sun, and people working in the bushes with clippers and spades suggested a peaceful backyard gardening scene. Only the mosquitoes said otherwise. Even with strong, high-DEET repellent, mosquitoes landed in hair and on the back of the neck and swarmed and buzzed inches from one's face. The sight of them was inescapable, hovering randomly in the field of vision like the stars that appear after a blow to the head. Most researchers wore mosquito shirts with hoods of mesh that covered the head like a beekeeper's suit. Combined with repellent, gloves, and thick pants and socks, a mosquito shirt prevented bites, but underneath the fabric the sun and the work combined to produce a slimy body film of mixed sweat and DEET. Sticky, but better than going mad.

But some scientists couldn't wear mosquito shirts because of their work. Henrik Wahren, a University of Alaska Fairbanks postdoc, and his research assistant, Amy Carroll, worked without covering day after day in conditions that I couldn't stand even for an hour. They had reached the point of seeming not to care. Their project was as mind-boggling as the pluck. One decade earlier, on a one-kilometer-square grid, a hundred evenly spaced, one-meter-square grids were precisely marked, and within each meter a hundred evenly spaced points were precisely marked, and at each of those exact points, a species of plant or fungus was recorded. Now Henrik and Amy were finding each of ten thousand exact points again and repeating the identifications. Even in this poor ecosystem, 250 species could be found growing, mostly scores of species of mosses and lichens that, to the untrained eye, looked identical. Henrik knew these species, but he often needed a hand lens to tell them apart, and he certainly couldn't do it through mosquito mesh. At each tiny point he identified a species by its Latin name and Amy, seated under a parasol to keep the sun off the screen of her laptop, typed a code for that species into a spreadsheet. The work went on like this, in hypnotic fashion, until Henrik said, "Ooh," at finding a certain variety of sphagnum moss under the magnifier. Sometimes, an identification of a sphagnum moss required taking a sample back to the lab and examining individual cells under a microscope. Some species of lichen required chemical analysis to tell them apart. Like Zen masters, Henrik and Amy continued their work without even swatting at the mosquitoes that surrounded them in clouds. They smiled indulgently at my manic attempts to escape the bugs, as if they had moved beyond that, to another plane of existence. In the margin of my notebook I wrote, "Incredible that people do this. Power of the scientific culture."

Henrik's belief in the work gave him perseverance. This project would help nail down ecological changes over the last ten years and help scale up from the single tundra plant to the landscape: when satellites looked at this square kilometer of ground, they would know exactly what they were looking at, and that would help adjust their vision. His true interest was in the landscape, not the twigs he stared at through his glass. Henrik subscribed to the winter theory of shrub growth. He said the loss of mosses due to drying would reduce the summer insulation of the ground, warming it and counteracting the cooling caused by the shadows of the shrubs. But he didn't expect narrow process studies would get to the heart of the matter, because the system was too complex. He, like several other scientists at Toolik, hungered to see ecology as an Alaska Native expert would—although at Toolik the scientists had no contact with Natives.

Henrik said it was a career imperative to publish, even stuff you didn't want to publish, and to read as much as possible—he monitored fifty journals—but that was not where the most important knowledge came from. "We've got the papers, but invariably you turn to the elder. The person who has been in the field and who has an intuitive sense. They notice things you don't see, and they pick up things you walk by. And they put it together into one holistic thing that blows your mind." After enough time in the field as a student, Henrik had felt that sense growing, apart from his training in identifying species of plants. "I finally started seeing things. Recognizing trees, recognizing rocks. And I'd say, 'I've seen that before, and there's something wrong with it.' Finally you start to see the landscape, and that was a revelation to me. It can never approach that of those who actually live in the landscape, but it moves you forward where you can really understand the landscape and not just see it through some kind of filter or system. You just look without identifying. That's always a joy, when you get to that point, where it all drops away, and you don't care what it's called."

I hiked away from Henrik and Amy to enjoy the landscape myself, letting the bugs fall behind me and feeling the slight breeze cool my sweat. After only a couple of days, Toolik was altering my sense of time. The camp reminded me of the sanitarium in *The Magic Mountain*, Thomas Mann's endless novel, which I read in translation long ago as an assignment and never finished and which was inextricably combined in my memory with Princeton's spring sunshine on the cemetery across from the screened porch where I sat reading for hours, smelling the blooming dogwood, losing the train of the abstract debates between Messrs. Settembrini and Naphta, who disputed through page after page of tiny type until time itself stretched and collapsed into a mental abstraction, which may have been Mann's point in the first place. Hiking over the tundra was easy despite tussocks the size of soccer balls. I rose over a rounded ridge, descended to a damp place, climbed another ridge, followed it along; I could gauge direction by the surrounding mountains, but the camp and the experiments quickly disappeared behind in the gracefully patterned repetition of the topography. Hiking on this dry tundra was like hiking on a continuous mountaintop. Distant points were almost stationary. Close up, the tundra came into focus and passed in the space of a few steps. There was no middle distance. Underfoot I came upon discarded caribou antlers, weather-scoured as smooth as finely finished wood, some with a bony surface like molded plastic, others patterned gray like decaying driftwood. As the sole definite features offered up by the smooth green landscape, they seemed to merit careful inspection.

The rules of the chariot races that night required each team of two—a male bike rider and a female luggage cart rider—to chug a beer before they could mount their bike and chariot and speed around the compound. Dirk had thought this out well: one of the bikes was clearly inferior to the other, so the drinking phase of the race established a fair criterion for which team would get the better wheels. A crowd gathered in front of the terrestrial ecology lab, cheering for riders who poured beer into their mouths, passed the bottle off to a partner for final gulps, threw the empty down, pedaled off furiously, tipped the luggage cart precariously on the turns while its occupant, clinging to the plywood sides, stayed low. Gus and Josh, the senior scientists present, watched with detached amusement. With the end of round one, Dirk declared the beer drinking part of the competition at an end—they had already finished off a box of his good craft brew, Alaskan Amber, which had to be imported by personal trips to Fairbanks—but the spectators wouldn't hear of it and more beer appeared from nowhere like loaves and fishes. In the later rounds, the young men pedaling the bikes, who had by now gulped down three or four bottles of beer in quick succession while exercising vigorously, were beginning to sway and look a little green. Gus said he was off for the sauna and soon he appeared walking across the chariot course with a towel wrapped around his waist, watching the chariots pass by before continuing to his destination. Toward the end, one of the luggage carts fell to pieces.

On the Magic Mountain there seemed to be an infinite number of ecology experiments to learn about, and the next day I looked at many more and listened to explanations of their amazing complexities, but my mind was filling up and giving way to lethargy. I took another long hike, then ended up on the sunny deck behind the dining hall absorbing the warmth, flipping through worn-out magazines, and carrying on irrelevant conversation, waiting for that evening's competition. Dirk had proposed bicycle jousting with PVC sewer pipes as lances.

That was when Ken Tape arrived. He came down out of the sky in a helicopter and strode into the dining hall with his father, Walt, and a pilot, looking for food. Matthew Sturm had missed the trip due to illness—he was experiencing bad headaches—but the other three had been in uninhabited bush country for several days. Dinner was over and put away, so they turned with gusto to the food set out on the snack table, building thick sandwiches of cold cuts, gathering piles of junk food, tucking in, shoulders bent down, like they really meant it. Ken's hair was standing up, as always, and he and Walt had grown madmen's ragged beards all over their faces, the fashion look of people who have been away from mirrors for a while. But they looked great. Ken's face was open with wilderness euphoria: he smiled a smile of revelation, looking over the top of people's heads, making quick, efficient movements tuned to outdoors work, away from society. Somehow, he very soon gathered three attractive young women around him to plan a future wilderness trip together.

I split off with Walt, a mathematician from the University of Alaska Fairbanks who normally studied and photographed sky phenomena such as halos around the moon and sun. We unloaded the helicopter together in the low-angle nighttime sun. He and Ken had covered half of Arctic Alaska from this midpoint to the west, taking scores of photos to replicate the old shrub pictures. The helicopter allowed slight adjustments in elevation, distance, and position to produce a view through Ken's large-format camera that exactly matched the prints from the old photographs. The work had gone better than expected and they were excited. They had mastered the difficult techniques of repeating the old photographs, and they had seen a lot of shrubs.

"We never see less," said Ken, who rejoined us. He was overflowing with the excitement of discovery. In three days of flying hundreds of miles about two hundred feet off the ground they had seen people only twice. "A lot of the time you're coming down a river and three bends in a row will look exactly the same. It's incredible flying over and seeing how regular everything is, how the same patterns are repeated again and again and again." In many places, permafrost effects carved sharp-cornered polygons on the tundra. "It looks man-made. Square lakes. I've flown over them on the coastal plain and I've flown over them here. I can't believe it.

"I just like the spaciousness of it."

I asked Ken a couple of times to compare his work, zooming over vast country, grabbing a fifty-year record of change with a flick of a camera's shutter, with the work at the plots at Toolik and the pluck, which was still going on, leaf by leaf, in one of the labs. He wouldn't do it. "No comment, man. No comment, no comment. Those people are hard-core."

Presently Ken and Walt had loaded their SUV and were ready for the drive back to Fairbanks. I had planned to stay longer, but the sun would be up all night and I felt restless to see Alaska speed by my window again. Waiting for the jousting competition, spending this spectacular, golden night that way, no longer seemed justified. It seemed like a waste. I thanked my hosts, cleared out my room, poured the gas from my rooftop cans into the tank, and flew away down the road, back over Atigun Pass, and through the blur of the landscape.

In 1858, Henry Adams, just out of Harvard, left his home in Quincy, Massachusetts, for a European tour. As the grandson and great-grandson of the sixth and second presidents, he had by birth a privileged and promising position in pastoral America. After arriving in Liverpool, however, he got a first look at the future, the dirty industrial works of Birmingham passing outside the window of his train. This was the physical manifestation of the coming mechanical age: "the plunge into darkness lurid with flames; the sense of unknown horror in this weird gloom which then existed nowhere else, and never had existed before, except in volcanic craters; the violent contrast between this dense, smoky, impenetrable darkness, and the soft green charm that one glided into, as one emerged—the revelation of an unknown society of the pit." On arrival in London, Adams also was dismayed by the closed, class-ridden quality of European society. His impression of the city: "heavy, clumsy, arrogant, purse-proud, but not cheap; insular but large; barely tolerant of an outside world, and absolutely self-confident."

That was also how New York City appeared to a visitor arriving from the invisible province of Alaska in the late twentieth century. I was born in New York but grew up in Alaska from age three, returning on annual East Coast trips to visit extended family and later to attend college. I remember on more than one arrival the overcharged vividness generated by excitement and jet lag. As you left the airport, the air would be damp and dirty—you could taste it on your tongue. Cars flung themselves along immense roads through corridors of billboards and wrecked factories, a nauseating visual surfeit under weird, unnatural light. Single complexes contained more people than my entire state. Coming from classless Alaska, I was intensely uncomfortable to be served in so many ways, by doormen, car parkers, gardeners, domestic help, even bathroom attendants—always brown people serving white people, and always with an exaggerated politeness that only accentuated the distance. With few exceptions, those jobs didn't exist in Alaska, and in the exceptions the people behaved as equals. Over four years of college I acclimatized to the muggy heat and the spatial and social narrowness of the megalopolis. I finally lost the claustrophobia, the sense of visual fatigue of everything being too close, the yearning to fly straight up into the sky to cleaner, cooler, drier air. But still the thought of living there permanently instead of in Alaska gave me a shiver of dread.

Alaska had both Natives and natives. The upper-case Natives were the indigenous people whose village culture had far less in common with that of Anchorage and Fairbanks than those cities had in common with Seattle or L.A. But the lower-case native Alaskans also had a culture. Not a universally shared culture—the urban gang members and transient British oil executives in Anchorage imported their own worldviews—but many did share values and ways of associating. They valued time spent outdoors, the ability to do things yourself, and individualism. You could be a weird-looking nonconformist, you could hold wacky political views, and you could do stupid, destructive things and still be an Alaskan. But when some prissy Anchorage subdivision adopted a covenant requiring residents to paint their houses only certain colors, that was un-Alaskan, and it created a storm of outrage in the local newspaper. In my Ivy League family, I've gone native more than anyone else. I'm almost second-generation Alaskan. I own a shotgun for bear protection, I have a large family and a large car, and we use a motorboat for wilderness camping. That alone would have killed my grandfather, who called motorboats "stinkpots" and whose greatest pleasure was to sail across Long Island Sound, pick up a mooring, and serve a round of cocktails. I surprised myself one evening at a community meeting when I argued against making people remove the old campers, overturned boats, and stacks of weathered lumber from their yards; I just didn't like the idea of how the neighborhood would look sanitized. This was Alaska, after all.

Matthew Sturm and his inventor colleague Jon Holmgren were essential Alaskans (there were a lot more of the type in Fairbanks than in Anchorage). Jon's shop, in a big shell of a metal building, stood on a gravel pad of gold mine tailings north of town next to the little sawmill operated by another science team member, Eric Pyne. Jon's life was organized with magnificent focus. He liked making things, so he had read books to find out how to be a machinist; he kept a shelf of them, and on his workbench he had a technical handbook of 2,500 pages of tiny type, which he had read through twice. He had gathered the equipment he needed secondhand, huge hydraulic machine tools weighing many tons, which were up to forty years old. For materials, he had a big junk pile outside. On the morning I visited with my older two children, Robin and Julia, Jon took a hunk of scrap steel and, before our eyes, working with several enormous tools, formed it into a precisely machined part for a scientific ice auger. He had just finished making a replacement transmission gear for a gold miner's worn-out tractor, but his biggest market was equipment for Arctic and Antarctic science, a lot of which had to be invented based on ideas scribbled on envelopes and talked out while slapping mosquitoes in the yard. Jon had patented some of this stuff, but he said he wouldn't license his patents to anyone else because making money wasn't his primary goal. "I like building stuff," he said. After the tour, Jon demonstrated a catapult he had built, on a medieval design, hurling a large boulder far across the gravel pad. Then he let Robin and Julia drive a cart propelled by a water well pump motor—top speed about four miles per hour—which he kept at the shop for his own kids, adopted African American twin boys.

Matthew lived nearby, up a hilly gravel road among rattling birch trees. The headaches that had kept him out of the helicopter with Ken were serious enough to require a brain scan, which didn't find any cause; ultimately, a tooth he had broken on the Arctic transect was found to be the culprit. In the meantime, however, Matthew got temporary relief by digging up his driveway to solve a problem with his well. He and his wife, Betsy, had built the house with their own hands on eight acres of hillside woods. The house was lovely and long and, like a colonial farmhouse, had the varied lines of progressive amendments, with a steeply pitched roof tucked with bedrooms and Betsy's study, which was reached by a ladder from the kitchen. Matthew also had built a huge shop and garage on the steep hillside, with the downhill side held far above the ground with supports of his own engineering. Parked in the driveway he kept a riverboat with a huge V-8 engine. The woods around the house contained a trailhead for a seven-mile backcountry skiing route. Matthew and Betsy's daughter, Skye, was on the high school ski team. Betsy was working on a pen-and-ink guide to identifying juvenile Alaska fishes, drawing exquisitely detailed fry and fingerlings sharper than photographs; they looked like blueprints for building fish. She and Matthew were preparing for eighty visitors from all over the country for the bar mitzvah of their son, Elias. They were also planning construction of a remote winter cabin on five acres, at a place that could be reached only after hours of skiing. They all loved winter that much—the science treks and Fairbanks ski trails weren't enough. And they loved freedom, since that was what going into the backcountry meant.

Henry Adams didn't go west. His privilege and his pessimism made going west appear superfluous to him from a young age. Instead, he stayed home and watched the comfortable, unified world into which he was born explode into modernity's complex and indefinite multiplicity. Through his experience of ferocious change in the second half of the nineteenth century, Adams became one of the first writers to express the alienation of the twentieth, a theme that, after his death, dominated a hundred years of literature and art until it became a cliché even in pop culture. He seemed to predict the future accurately from one of his laws of history, his idea that mankind's exponential increase in power drawn from fossil fuels mirrored a similar expansion in the complexity of science and mathematics, which threatened to make reality unintelligible. In 1905, he wrote, science already had "betrayed the inadequacy of old implements of thought." Of the twentieth-century American, Adams wrote:

> For this new creation, born since 1900, a historian asked no longer to be teacher or even friend; he asked only to be a pupil, and promised to be docile, for once, even though trodden under foot; for he could see that the new American—the child of incalculable coal-power, chemical power, electric power, and radiating energy, as well as of new forces yet undetermined—must be a sort of God compared with any former creation of nature. At the rate of progress since 1800, every American who lived into the year 2000 would know how to control unlimited power. He would think in complexities unimaginable to an earlier mind. He would deal with problems altogether beyond the range of earlier society.

One could read the prediction as positive, but it wasn't. It was more like the plaintive cry of my parents when they asked me to come fix their VCR.

In the fall of 2002, I got on an airplane one night to fly from Anchorage to Washington, D.C. The first leg, from Anchorage to Seattle, had always felt like a step over a threshold to the faster, more man-made environment. That step has shrunk over the decades as communication and the spread of corporate homogeneity unified Anchorage with the national market, extending the plastic shell Richard Glenn called "mall culture." But even so, Anchorage remained a small town with pretensions. Wild land was never more than thirty minutes away, and the security checkpoint at the airport was quick and friendly. I usually saw people there I knew. On the plane I settled into my synthetic little personal space with my laptop to take advantage of the hours for work, harnessing the power and technology Henry Adams had imagined for the human gods of the future to nothing more profound than writing a formulaic travel guidebook that I owed to a publisher. Others flying over the dark Alaska mountains were using their computers to watch DVDs of formulaic Hollywood movies. The multiplying complexities Adams found so frightening turned out to be horribly mundane.

At Washington's Dulles Airport a distinguished white-haired gentleman sat in a metal folding chair, trapped. The security people had lost his shoes. Travelers struggled with their baggage from the claim area across a confusing ad hoc zone of plywood barricades and walkways to a sidewalk for boarding buses to the rental car agencies. Scowling guards peered suspiciously at each party, apparently barely able to tolerate anyone who would wait for a Hertz bus. After enduring all the waiting and crowding and institutional insolence, drivers took off toward the city at breakneck speed—or so it seemed to a rube like me—as if they were seeking to defy the airport's indignities by becoming their own unthinking mechanical juggernaut. At the other end of the freeway, on the Potomac River, ramps into the city split like strands from a frayed steel cable, cars slipping away as fast as water droplets. I ended up on the wrong strand, a conduit that trapped me into driving miles in the wrong direction. I cursed the place. Barrow's collection of square plywood houses on the cold and featureless Arctic plain seemed warm and inviting in comparison. I thought, "No wonder everyone here is alienated." At a gas station, the customers seemed to be cowering around pumps off to the side, leaving the front abandoned. Then I remembered: there was a sniper preying on the city. The newspaper carried the stories of random victims and the poignant details of everyday life cut off. Every story could have run with the same headline: "Next time, it could be you." At the hotel, all the guests were white like me and all the staff were smiling, nodding Indians. I smiled and nodded back like a demented bird. One of them took my car away—apparently, it was impossible for me to park it myself—and I gave him money, not knowing if it was the right amount, and wondered where the car might have gone, with the pencils and notebooks and reading matter I needed, but lacking the heart to ask the nice man to bring it back again.

My purpose in coming to Washington was to meet with the people who ran Arctic science, but I ended up focusing on the new building Glenn Sheehan and Richard Glenn were trying to get for BASC in Barrow. The previous spring, Ted Stevens had amended the energy bill with funding for the building and portions of the climate change legislation originated by Robert Byrd. That bill went to a House-Senate conference committee, which continued to drag on that fall. Ultimately, it died, but in the fall the negotiations were still a topic of speculation. A reporter told me the Republicans were treating the climate change issues as trading stock—something the other side wanted and they didn't—giving those provisions little chance of survival. Meanwhile, the science establishment opposed the building. It had become another front in the long war to stop Ted Stevens from funding his own science priorities rather than the priorities arising from the NSF's process. The NSF's Arctic officials talked about the BASC building as if it were a crazy idea. "If you put this big thing up there, how the hell do you run it?" said Tom Pyle, head of Arctic programs. "How much science is there that really needs it?" He preferred to put new money at Toolik Lake.

After my visits to both places, I couldn't make that preference compute. Toolik already had superb facilities for researchers concentrating on tundra, lakes, and streams in summer; while it was physically possible to work there in the winter, it would be a hard assignment that few had chosen to accept. Barrow offered pathetic facilities to researchers studying the land and the ocean, and the wildlife of each, and the sea ice, the snow, and their albedo effects, and the people of the Arctic and how they related to the environment and change. And studying over the winter was easy: pizza delivery was only a phone call away. The worn-out Barrow facilities were heavily used, and Glenn Sheehan believed that many more scientists stayed away because they didn't want to deal with the inferior equipment.

The subtle forces of the science funding process were driving Toolik's primacy. Working scientists lived on grant competitions, which they saw as a zero-sum game. Toolik was winning fair and square. A new building in

Barrow, with funding from outside the process, would be cheating. The money for the building itself might not take away from other grants, but the existence of the building would give Barrow researchers an unfair advantage. The opposition of science administrators was more complex and, from their perspective, more principled. They opposed the building largely because Ted Stevens supported it. Government administrators wouldn't say that, but some of the science elders in their peer group did, and that was how Ted saw it, too. He had shown his interest in climate change by holding Senate Appropriations Committee hearings on the topic in Fairbanks, for which he hauled north the heads of most of the national science agencies to testify before a committee of one—himself alone. To fund Arctic scientists' biggest dreams was easily within his power. But for the agencies to give in to any of the projects Ted wanted himself without the prior blessing of the science establishment would be to surrender objective merit to pork barrel politics, which was unacceptable regardless of the potential gain.

At least that was a prevailing view. To me, it looked more like a fight for power than for ethics. The competitive process of science was certainly better informed on this subject than Ted Stevens's decision making, but it was no less political and it wouldn't necessarily always yield better results. Syun Akasofu was right: scientists often held on to their own theories and their own ways of doing things for their own reasons, not because they had an insight into the truth. Besides, Syun said, the high level of competition for grant funding had made the process arbitrary by necessity. When anonymous peer reviewers and administrators received many more proposals of high merit than they could fund, they had to pick on some other basis. Now the black box of science decision making was putting out dubious results in the choice of Toolik versus Barrow. The choice looked like a cultural preference similar to the bias for a summer theory of Arctic shrubs because scientists visited only in the summer. The in crowd was at Toolik, so Toolik had more merit. Or so I suspected, without a way to prove it.

Ironically, the same scientists who defended the pure process that yielded funding for Toolik over Barrow joined with Barrow scientists in accusing the NSF of politics in its preference for Antarctica over the Arctic. The NSF was spending more than three times as much in Antarctica annually and was building a new South Pole station for $133 million. Except for the issue of glacier stability, Antarctica was less relevant to climate change research than the Arctic, because Antarctica was white year-round—albedo feedback was much less an issue and plants and animals were too scarce to process much carbon—and as a consequence both the predicted and the measured warming in Antarctica was slight. Arctic scientists consistently grumbled that the Antarctic money was mostly just to keep the U.S. flag planted down there. "Everyone went to Antarctica for reasons that were supposed to be scientific, but everyone knows it was political," said John Kelley. "Politics really ruled there."

John was a former director of NARL, former director of the U.S. project that successfully drilled Greenland's ice sheet to bedrock, and originator of CO_2 measurements at Barrow not long after the very first started in Hawaii—his work started the station that became NOAA's Climate Monitoring and Diagnostics Lab in Barrow. John also had spent two years in the 1970s as a program director at NSF, and he was as cynical about the new BASC building's chances as anyone. He predicted that the Ted Stevens money would be siphoned away for the automated CMDL operation while the NSF put its real attention at Toolik and in Greenland. And if the money did go to BASC, it wouldn't necessarily be good for Barrow, because ongoing support could disappear overnight without constant political vigilance.

"Only recently has NSF shown any willingness to go up there, and it has done so with great reluctance. There have been endless reviews and meetings," John said. "As soon as the interest changes, so will the money. It will fly out of there as fast as a speeding bullet, and what will you do then?

"I look at NSF as being just one giant superbureaucracy, with the people in there like Russian apparatchiks with one goal—cover our path."

I assumed that wasn't fair, but I didn't hope to know the truth of the matter. As a machine for spending money, Washington was as complex and inscrutable as an ecological system. Measuring its inputs and outputs might be relatively simple, but figuring out why money flowed in one direction and not another could involve processes with variables too numerous to analyze. In the end, each organism took away the resources its competitive advantages allowed it to win. Pure scientific merit might be one such advantage, but, if so, its competitive prowess was not great. Nonmilitary research and development accounted for only about 2.5 percent of federal spending, and the entire NSF budget was only 10 percent of the civilian R&D figure, and work on the Arctic was only 1 percent of the NSF budget. "We are scientists," Tom Pyle insisted, rather sadly. But he added, "I assume that if Ted Stevens wants it, it will happen."

Going to the NSF headquarters in Arlington, Virginia, to meet Tom meant penetrating two levels of security from the grand public atrium to the anonymous little offices where people held their meetings and wrote their e-mails. Tom, an older man, seemed tired. His shoulders slumped as we walked back to the office after an interview over lunch. He wondered out loud why the Alaskans couldn't work through the system. For example, why had that spring's conflict over the icebreaker cruise become political? That topic brought us to whaling, and then he mentioned he had received an e-mail that morning saying one of the whalers in Barrow had been killed in an accident.

As soon as we parted, I sat on the street in front of the NSF building and called Richard Glenn's home phone number on my cell phone. It was early in the morning Barrow time. Richard answered with a sad voice.

Malik was the one who had died, the wise old whaler and teacher, the master of climbing on top of the whale and delivering the fatal blow with the lance, the unerring follower of the whale. The one whose name I had not deserved. Before coming east, I had returned from a fall whaling trip with Richard and Roy, and I had seen Malik again. During fall whaling, when there was no ice near shore, Barrow crews hunted from boats with large outboard motors. On a windy, foggy day when most crews stayed in, Savik Boat received a call for help from Malik on the VHF and we flew over the water to him. Malik's boat was anchored off Plover Point, bobbing in sizable waves, just a skiff of riveted aluminum sides with a plywood forepeak to break the wind and spray, and down in the middle of it, Malik, smiling up at us with a face beyond cheerfulness, surrounded by his crew and family. In the wind and cold dampness the crews connected jumper cables from Savik's fancy new engine to Malik's old, worn outboard and exchanged laughter and warmth about the predicament and the meeting. Malik's engine started right up.

Richard said that on the day of the accident crews had killed three whales and Malik's boat (a borrowed boat different from the one I had seen) was in line towing one of them back to the beach. The Iñupiat pull the whale with a long, thick hawser to which each boat attaches with two lines, one at the bow and one at the stern. The danger comes if a boat's bow line breaks, because then the force of all the boats pulling can jerk the stern line, spinning the boat sideways and dragging it into the waves. On the day of the accident, the seas were up, but not unreasonably so in Richard's opinion. Rounding Point Barrow, where wrapping waves built on top of each other, the boats ahead of Malik's made a turn. A wave crested higher than his bow line, which pulled the bow under the crest, swamping the boat. With the boat full of water the pressure on the bow line was too great and it snapped. The stern line spun the boat backward and it quickly flipped. The rest of the crew came up safely and were pulled from the water, but Malik was trapped underneath the boat for ten minutes until rescuers pulled him out with a gaff.

Richard had been towing another whale. He saw the ambulances on the beach. "I went to the hospital and kept a vigil," he said. Most whalers stayed behind. The harvested whales were lifted out of the water with heavy equipment and deposited on the metal NARL runway for butchering, as was customary in fall whaling. The work could not stop or the meat would be wasted. Friends came from the hospital with the news, bringing prayers to the site of the butchering.

The funeral at the Barrow High School gym was the largest seen in Barrow in many years, "a real whaler's funeral," Richard said. Every crew that Malik had taught and for whom he had thrown the harpoon or killed a whale wanted to help carry his body. It seemed he had taught everyone in town, not just members of Savik Crew. The weather was cold and the wind was blowing up to forty-three miles per hour, but the series of pallbearers taking turns formed a long parade as they carried the casket three and a half miles over gravel and thin strands of snow to the cemetery.

When I talked to Richard on the cell phone, he was waiting to go out whaling again as soon as the weather settled down. Every day each crew had to make its own decision, balancing the size of the waves—which were large that year—with its desire to land a whale. There was no one to blame for Malik's death except the inherent danger of whaling from small open boats in the rough autumn waters off Point Barrow.

I told Richard some of what I had learned about funding for the new BASC building, but he already knew everyone I had talked to and more about the politics than I could relate. I mentioned the street where I was sitting in Arlington, and he gave me directions to the nearest Metro station and told me which train to catch back into D.C.

CHAPTER NINE

The Spirit

FROM THEIR COZY MOUNTAIN HOME in the woods of western North Carolina, April and Steve Cheuvront could walk among tall, misty trees into wilderness, and often did. Inside, the evening was quiet—most of the neighbors were summer people and the nearest major road was far off—and while April prepared venison lasagna, Steve showed me pictures of their recent vacation to Fiji. After her Arctic transect, he had chosen their summer trip, a week on a tiny, remote island where they slept in a grass hut at night and fished all day in a little boat with a native guide. April thought Steve would fall in love with Alaska, given his enthusiasm for hunting and fishing, but at the suggestion he only smiled with gentlemanly reserve. He was a civilized, southern sportsman, a black-powder hunter, a stream fisherman. His one-man country law practice had brought him payment in the form of new tires for the car and a banjo, leaning against the wall, that he didn't know how to play. He liked that; he fit here. Steve told me quietly, when we were apart from April, that she would never really move to Alaska, so far from her parents in West Virginia. Soon he and April would be ready to build a house and start a family on the five acres they had bought in the mountains. That version of the future, which predated talk of Alaska, remained Steve's ideal, and in the quiet coolness of the evening it seemed hard to improve upon.

In the morning, April and Steve got up with the sun to drive the winding road down the mountain to Morganton. North Carolina was having a hard time, with many low-skilled manufacturing jobs disappearing—people blamed free trade—and the town, strung out on a long main road, seemed faded and tired. April's middle school was in a modern building, but it was crowded and many of the children came from poor families with problems at home. To me the students' days seemed unbearably narrow and sedentary, devoid of freedom. At twelve, thirteen, and fourteen, ages when the body and mind explode with wild energy, they marched single file in silence from class to class, with no time spent outside and no unstructured minutes through a long day. I showed my photographs of whaling in Barrow to two of April's classes. I introduced the members of Savik Crew and asked the children to imagine learning to throw a harpoon like Eben and Benny, boys not much older than they, who were helping provide for their entire community. I told them about the courage required to hunt a whale and of Malik's recent death. April said I was like a guest speaker from another planet—most of the children had never traveled far beyond the immediate area of Morganton, and despite the farms in the area, most of them had no understanding of where their food came from.

They were polite audiences, but many of the students, especially the older ones, crossed their arms and made it clear with their questions that they thought whaling, even for food, was wrong. After one of the talks, April asked them to write brief reaction papers. A few boys from hunting families envied the whalers. "Not only the hunting is exciting, but the endless amounts of land and snow, like a million pillows lined up on the ground," one wrote. "The fact that men can kill a huge animal and feed themselves is amazing to me." But more wrote comments such as: "I wouldn't take part in killing the whales. I would probably be out helping sick or hurt animals instead." Several found the whole thing disgusting. "It would make me sick to see the whale being cut open, so I would have to leave," a boy wrote. "I don't think I would want to go outside very much because of all the polar bears roaming freely around everything." A girl thought whale hunting was wrong and sad and, like several others, thought the Arctic was depressing and horrible. "I would hate it up there. It would be cold. I hate cold weather. My life is complicated enough. It would be more complicated in the Arctic."

The Iñupiat were used to these reactions. At times, they seemed constantly embattled. That spring the International Whaling Commission had banned Eskimo whaling entirely. In May, soon after the ice in Barrow broke off and the whalers floated out, George and Maggie Ahmaogak left for Shimonoseki, Japan, to lead the Iñupiat delegation at the IWC meeting considering renewal of their quota. The scientific team, including Craig George, had already presented their evidence that the bowhead population was strong and justified a stable quota. But the big issue of the conference was the effort by the Japanese delegation to end the 1986 moratorium on commercial whaling. New member nations had joined the IWC in the Japanese column amid allegations that their votes were bought by Japanese foreign aid. Critics said there was no other explanation for Mongolia's sudden interest. The Japanese countered that environmentalists had recruited antiwhaling nations into the IWC in earlier years. In any event, the Americans and their antiwhaling allies won narrow majorities on key votes, blocking Japanese maneuvers toward ending the moratorium—even though one U.S. diplomat later admitted to me being privately sympathetic to Japan. She saw no reason the Japanese shouldn't take a commercial quota of 50 minke whales from a population estimated at 750,000 to 1,000,000 animals. The objections were cultural, not scientific. The British fisheries minister said his country would oppose commercial whaling no matter the numbers because whaling was "unnecessary and cruel." The suggestion that hunting was cruel especially offended the Iñupiat; as I heard one say, "At least we don't keep animals in little cages."

The vote on the new five-year quota for the Iñupiat came up a few days later, with a three-quarters majority needed for approval. U.S. diplomats told the Iñupiaq representatives to keep cool because trouble was expected, and they did maintain their dignity as a midlevel Japanese environmental bureaucrat, Masayuki Komatsu, seized the microphone at the plenary session and called for a rejection of the Eskimo's quota to expose American hypocrisy over the commercial whaling issue. Japan won the vote, then won a reconsideration, holding the one-quarter of delegates it needed. Masayuki Komatsu became a sudden star in Japan, where TV viewers had seen his performance and flooded his ministry with positive e-mail. That may have helped motivate Japan to stand firm through the summer, despite U.S. diplomatic pressure and international condemnation in the media, even as George Ahmaogak made clear what everyone in Barrow knew: the Iñupiat would go on whaling whatever the IWC said. The British charged that this was why Japan had defeated the quota—to force the Americans to defy the IWC and thereby destroy it.

Maggie had no sympathy for the Japanese. She said they hadn't done their homework to prove their need for whaling or to conclusively establish the whales' population, as the Iñupiat had done. Besides, Japanese whalers hunted for profit, with its incentive for excess. Traditional Iñupiaq whalers wanted no more than their villages could use. Organizations dedicated to whale protection recognized that difference and acquiesced to the Eskimo quota while fighting Japanese whaling. Nonetheless, some in Barrow, including Craig George, blamed the antiwhaling groups in large part for the Iñupiat defeat in Shimonoseki. Despite the differences, they said Japan had a good point. The U.S. stance was like India rounding up votes to block Americans from eating beef. Craig said management of minke whales was so conservative that if the same principles were applied to fish there would be no fishing anywhere on the globe. The only justification for the ban on killing a tiny percentage of the minke whale population was that people liked whales; Americans, Europeans, and their supporters on the IWC simply didn't want the Japanese to eat such popular animals.

Everyone was guilty of these prejudices. In Shimonoseki, Craig and his companions made friends with a couple who ran a sushi restaurant where they often ate. At first, the Japanese couple avoided conversation with the Americans, possibly because of the heated political situation in town, but Craig's group told them, "We're OK, we're from Alaska and we hunt whales." After that the couple opened up and the two groups shared clumsy, phrase book Japanese and English. Then Craig put together the Japanese phrase "Caribou are delicious," and the conversation came to a grinding halt. "They were shocked," Craig said. "One of the interpreters said, 'You know, reindeer is not something that is suitable for eating.' That was a real faux pas. I could see the shock in their faces. How on earth could you eat these animals? It was like eating whales in this country."

For almost any animal you choose, you can find someone in the world who relishes eating it and someone else who finds the idea revolting. Yet many Americans and Europeans behave as if there were a rational basis for a bias against eating whales. Whales are large, probably intelligent, and live a long time. But no hierarchy of moral edibility holds for other animals based on those criteria. Bison are large. Pigs are smart, and so are octopuses. Sturgeon, the source of caviar, live as long as whales. It is possible to come up with a rational hierarchy to explain which animals one can ethically eat, but avoiding hypocrisy can take you places you don't want to go. Utilitarian philosopher Peter Singer has pointed out that many animals are more aware than human babies and could have a greater capacity to feel pain and loss. He reasoned three levels of a right to life: on the top level, self-conscious beings, such as humans more than a few months old, apes, whales, dolphins, monkeys, dogs, cats, pigs, seals, bears, and likely others, which can never be killed without their consent; at the next level, beings who experience suffering but cannot think about themselves or their future, including babies, severely mentally disabled people, and lower animals, which can be killed but not exposed to unnecessary suffering; and at the bottom level, everything else, with no ethical claim on existence. This reasoning could as easily support eating babies as not eating animals (Singer has condoned killing severely disabled children), but Singer's arguments are ultimately based on preferences, so he became a strict vegetarian instead.

Singer's philosophical promotion of animals to human status started the animal rights movement. In 2003, People for the Ethical Treatment of Animals launched an advertising campaign that equated chicken coops with Nazi death camps. Their posters showed rows of starving Jews next to rows of caged chickens with the headline TO ANIMALS, ALL PEOPLE ARE NAZIS. PETA members at least had a logically consistent reason for opposition to whaling and hunting, although they couldn't make a moral argument to anyone unwilling to equate people and animals.

But why shouldn't they be equated? Shirley Tilghman, the president of Princeton University and a leader in the human genome project, said the project's most profound finding was the genetic similarity of people and other animals: we share 95 percent of our DNA with other primates, 90 percent with mice. There is no uniquely human gene; chimps have them all. When I asked, during the question and answer portion of her lecture, what that should tell us about our relation to animals, she joked that I should ask Peter Singer, who usually attended those sessions, and didn't really answer the question. Like many, she appeared unready to deal with the feeling that eating creatures with a genetic code so like our own seemed vaguely cannibalistic.

At least so it would seem if our kinship with animals meant that animals were human rather than that humans were animals. No one has suggested that polar bears are immoral for hunting seals or has called on the IWC to deny orca whales a quota of whale meat because of their similarity to their prey. Few mammals treat members of other species as equal in value to their own offspring. (I doubt any people do, either.) Living organisms seek to pass on their genes. They compete for resources with other species and with other individuals of their own species to make sure their line survives. If that means eating somebody, fine.

Animals cannot give one another rights, because rights are a human concept. Indeed, rights, whatever their source, function to preserve humanity from the brutal state of nature. Our social contract allows us to cooperate in a community; the community is one of the competitive advantages we have over other species, allowing us to adapt and thrive in almost any habitat. Rights help cement communities by providing universal justification for the rules implicit in our social contract. Within a human community, although not always beyond it, the right to life helps to keep people from killing each other in situations where animals of other species would not refrain. If rights were reciprocal, like a contract, then a decision to give them to animals would lack moral force because animals cannot return the favor.

But if rights were simply a deal made on a handshake, no one would want to give them to animals. Rights have power because of their transcendent, spiritual appeal, an authority that demands obedience regardless of the circumstances. In Western culture, the name of that universal authority is God. In his history of Western civilization's relationship to nature, *Wilderness & the American Mind*, Roderick Frazier Nash starts the story with Genesis, in which God gives humanity dominion over the earth and the animals, moral authority to grow crops and domesticate livestock. People resemble God, with eternal souls, and animals do not; case closed. For most of history, the circle of civilization contained tame land and animals that served people and were essentially good. Outside the circle, wilderness began, and it was bad and in need of control.

In the eighteenth and nineteenth centuries, however, Romantic philosophers and writers began reversing the negative, characterizing the pure nature outside the town gates as the good. In the twentieth century, with the rise of industry and man-made complexity critiqued by Henry Adams, this reversal of good and bad became the main cultural current. By the end of the century, the pop culture of alienation defined the content inside civilization's walls as an absolute evil, a degenerate concrete jungle and cyber wilderness against which the only defense was disengagement and irony.

God didn't leave the picture, however, He just changed sides from civilization to wilderness. Nature writer John Muir helped put Him there with ecstatic prose that found a church in every other grove of trees. The idea originated in the New England transcendentalist movement of the 1850s, but Muir carried it west and added an essential new element: to get in touch with God in the woods, it wasn't enough to gaze at placid Walden Pond; you had to go out into raw wilderness and be really uncomfortable. Wilderness became the setting for spiritual purification. A century later, although the explicit reference to God was usually gone, a large subculture of wilderness users believed that an authentic outdoors experience must be physically challenging and ascetic, unsullied by any manifestation of civilization. Mountain climbers and the like were made heroes in proportion to the danger and hardship they brought upon themselves. The spiritual epiphany of wilderness became such a cultural commonplace that it was sold in travel brochures.

The National Park Service was founded early in the twentieth century with a traditional outlook on recreation, but by the 1970s it had become the Vatican of the ascetic wilderness ethic. By the end of the century its focus had shifted to ridding parks of the facilities built earlier, in order to return the land to the pure state that existed before human intrusion. Little action was needed to attain that goal in the Alaskan parks, but the parks in the Lower 48 were too small and too disturbed to revert truly to the natural ecology that existed prior to the arrival of settlers, so park service personnel made adjustments to maintain the desired effect. On Yellowstone's northern edge, rangers drove bison back within park boundaries, while on its southern edge at Jackson Hole, elk were fed bales of hay to get them through the winter. Rocky Mountain National Park had a bighorn crossing guard posted on a highway. Great Smoky mowed the mountains bald to keep them from growing over. High hunting-bag limits just outside various parks helped keep ungulates from stripping the vegetation inside the boundaries.

While the park service stage-managed nature to represent the pure state it had selected, rangers also screened the audience to prevent disruptions during the performance. To venture into the backcountry at most parks, visitors were required to obtain a rationed permit and view a video teaching them how to camp without leaving a trace; at some parks, back-country travelers even had to carry out their feces. In light of the small size and heavy use of the Lower 48 parks, most of these measures were necessary to maintain the wilderness illusion. But the implied moral weight behind them was much stronger than courtesy. I began asking backpackers if they would bring an AM/FM radio on a national park trip. They would recoil as if I had suggested carrying a boom box into a cathedral.

None of this made any sense to the Iñupiat. When they camped, they brought along as much of home as possible; when they camped, they were home. They carried a radio to keep in touch. They did leave a trace: man-made landmarks were valuable for figuring out where you were. Nash points out that the word *wilderness* doesn't mean much to indigenous people, because it is a concept defined as separateness from civilization. One strain of thinking in "wilderness ethics" has held, therefore, that Native Americans living in nature cannot appreciate the meaning and value of wilderness as well as city people. By the same reasoning, only submarine crews can appreciate the sky, religious self-flagellants know best the value of nerve endings, and it takes an illiterate to get the good out of a library.

Richard Glenn told me he had taken care of the "Patagonia crowd" on Arctic trips. (I quickly checked to see if the Patagonia label on my fleece jacket was showing.) These urban people had to be watched carefully because, in the Arctic climate, their urge for noble suffering could kill them. The point, Richard said, was survival. "It's not a game, and it's not a guilt push. To me, those kind of people, they must live in excess, they must live against this, so they have to go the other way to even themselves out. They live so much on pavement, in automobiles somewhere, that they have got to do some spartan purgatory so that they can feel better. But even still, the thing that they buy—that little tiny stove and that little tiny tent and all that—it's top of the line, and then it becomes competition in that."

From the Iñupiat point of view, such guilt-ridden people—with their perverse, puritan conception of the outdoors—were the ones who wanted to stop the Japanese from eating whales and to keep Alaskans from taking oil out of the ground. They were simply cultural imperialists trying to impose their values on others. Some environmentalists' own spiritual language betrayed them: Senator Joseph Lieberman, like many others, said ANWR oil drilling would be a "desecration." But, leaving out religion, I could see both sides. Environmentalists believed that the tentacles of civilization had reached far enough, and it was hard to disagree. Untouched land had value simply based on its scarcity. An Audubon Society official said, "People around the country get enormous satisfaction from knowing that there is land where humans aren't present, that there are blank spaces. And oil drilling is incompatible with that feeling."

As understandable as such feelings are, however, they amount to preferences that could change from day to day depending on popular tastes and economic needs. One day, people around the country might decide that the conceptual purity of ANWR isn't worth leaving energy stranded there. Thoughtful wilderness advocates began grappling with this problem decades ago in other development fights. They turned to philosophy for more absolute justifications that could raise preservation above the economic marketplace and do it without explicit application to God. Drawing on the writings of Aldo Leopold, Roderick Frazier Nash tried to give rights to nature, stretching the circle to include even rocks. He suggested that strip mining raped land literally, not only figuratively. Violating nature, like violating a woman, was not a question of economic price but simply wrong. Even as he advanced the point, however, Nash had to admit that exactly what it meant wasn't clear—if strip mining is rape, is digging in your garden an unwanted grope? Essentially, nature's rights depended on setting an acceptable level of human use and denial. "Keeping human hands off of some environments demonstrates an encouraging capacity for self-restraint on the part of a species notorious for its excesses and greed," Nash wrote. But if our species is notorious for excess and greed, it is only among certain of its own members, or perhaps God. Other species neither judge us nor share our capacity to conserve resources for future generations, much less to leave stores of energy idle in order to avoid potential harm to "the intrinsic rights of other species and of ecological processes."

There is something suspect about this promiscuous distribution of rights. It seems too fast and too easy. Rights should come from deeper, down in the gut, not from the capricious intellect, which constantly offers up justification for one's own preferences and prejudices. Peter Singer, an atheist, defines the dawning of a person's unique right to life as the moment more than a month after birth when he or she develops a consciousness that perceives itself as existing over time. Consciousness, the thinking "I" that seems to control our free will from moment to moment, has been a popular candidate for the self and the soul through the history of Western thought. It gets to write the books, after all; it's a rather self-important little part of the mind. But consciousness is unnecessary and sometimes counterproductive for thinking, learning, and reasoning; one psychologist theorizes that consciousness is essentially a metaphor, a learned way of mentally modeling ourselves in the world. Consciousness often fails us when we need to synthesize large amounts of information, or produce creative insights, or make difficult judgments. In five acts, *Hamlet* demonstrates the pitfall of self-conscious thought, the torment of thinking but not knowing.

Knowing often comes from somewhere else. Farmers knew they had God's gift of dominion over nature. Eskimos knew they shared membership in nature with the animals they hunted. Beliefs about nature that were created in the twentieth century, with rights built up on an intellectual scaffolding, developed instead from a culture without a direct relationship to nature at all, despite national park vacations and dreams of blank space on a map of Alaska. In this culture, nature was a concept, sitting alongside the idea of cyberspace. For some of the children I met in North Carolina, nature was simultaneously a fuzzy ideal and a terrifying reality. They were meat eaters revolted by the thought of the killing and butchering of animals. They knew Porky Pig and they knew baked ham, but nothing in between. I doubted intellect could bridge that chasm.

Knowing how one's own physical being fits in the world takes experience in reality, learning from nature itself without an intermediary ideology or teacher, from living, seeing, and surviving—the experience of being an animal yourself. That knowledge, I believe, represents a more essential me. Contrary to Peter Singer, I would say the flittering "I" of consciousness is not what makes me a person.

On certain spring days in the mid-1990s, clouds of spruce bark beetles took flight among the big spruce trees around Kachemak Bay, 120 miles south of Anchorage. They could be seen from miles away, rolling down the Anchor River valley. People who witnessed the arrival sometimes felt like they were in a horror film, the air thick with beetles landing in their eyes and catching in their hair, and knew when it happened that their trees were destined to turn red and die. My wife's mother lived in Homer, the main town on the bay, and all our lives we had walked on pebble beaches and inspected tide pools there, and hiked in the woods, and fished and dug clams and camped out. As we drove the 230-mile highway to Homer in the fall darkness those years, bonfires as big as houses lit the night, showing flashing glimpses, as we passed, of heavy equipment and human silhouettes tending the fires, and new fields of uneven ground with big stumps protruding like the tips of broken bones.

The Lutz spruce, which grew few other places than around Kachemak Bay and north across the Kenai Peninsula toward Anchorage, were tall, symmetrical trees with long, sagging bows that muffled sound and stilled the wind while allowing plenty of sky to show through. In the winter and early summer before the brush grew too high, you could ski or walk through mature woods without a trail, passing from clearing to shadow and back to clearing, imagining the enchanted forest went on forever. The Lutz was a natural hybrid of the moisture-loving Sitka spruce, the giant, shaggy tree of the temperate rainforest that extended from Oregon to here, and the white spruce, a hardier, crisper, smaller tree that thrived along with birch in the dry, cold boreal forest from here north to the Arctic. As you traveled northward from the southern side of the bay to Anchorage, crossing the gradations from the damp marine climate toward drier inland skies, the trees changed in small gradations, too, with white spruce genetics gaining on Sitka spruce ancestry mile by mile. Nature had tuned each stand of trees as if reading the lines on a precipitation or temperature map.

Ed Berg arrived in Homer in 1977, a member of the Vietnam generation. As a student in the 1960s, Ed had worked on a Ph.D. in geophysics from the University of Wisconsin in Madison but stopped short when he came to believe the government might use his knowledge of the earth's crust for militaristic purposes. He got a Ph.D. in philosophy instead and became a carpenter. In Homer, he finished the shell of a home down a narrow, muddy road among towering spruce trees near other people like him. Homer was a haven for the world's displaced nonconformists. Dirt roads, trails, and ski tracks led through the woods to fanciful and half-finished self-built homes on large tracts of land owned by people without much money—not only ex-hippies and war protesters but also modern pioneers and colonies of Russian Old Believers evading Peter the Great's church reforms. Ed's next-door neighbors, far out of sight in the trees, were an outboard motor repairman who worked from home and his wife, who took to her deck each morning to sing in a grove of high trees that reminded her of a cathedral.

Entomologists thought Kachemak Bay was immune from the spruce bark beetle, which was generally not a very aggressive pest. The beetles' two-year life cycle required a combination of favorable conditions for an outbreak to get started. Adults flew to find new trees as hosts in the spring when the air temperature was at least 60 degrees F, then burrowed in to spend the rest of their lives in the phloem, the inner layer of bark that carries food created in the needles down to the tree's roots. If the temperature in spring was too cool, the beetles couldn't fly. If the trees were full of flowing resin, the beetles were entombed in their burrows before they could reproduce. But if warm weather came early in spring when the ground was still frozen and tree roots unable to suck up moisture, the beetles could fly and find vulnerable dry trees everywhere.

Every few decades the region had exceptional stretches of weather that would allow the beetles to infest an area, but the return of cool, damp conditions would knock down outbreaks before they could spread far. In 1987, however, a string of warm, dry summers began, and it never broke. Several summers were so long and warm that the beetles completed their life cycle in one year instead of two, doubling the hit on the forest. They were so numerous they wiped out young, healthy trees as well as old ones, leaving some stands without seed stock for regrowth. The beetles were already getting started in 1988 when Ed Berg went back to school to get a Ph.D. in botany. When he returned in 1993 and went to work for the U.S. Fish and Wildlife Service, he could see the trees in his neighborhood were bound to die.

As dead, red trees spread, and one community after another mourned its forest, Ed turned to low-tech tree ring study and statistical analysis to understand what was happening. He and his colleagues established that the warm weather was unprecedented in 350 years and the regionwide beetle epidemic unique for as long as the tree ring record could tell—well over 250 years. The regional lines of climate gradation to which these trees had adapted so precisely had jumped and made this a far more hospitable habitat for beetles. Four million acres of spruce died, the largest single insect kill of trees in North American history. The beetles stopped only after more than a dozen years, when the available timber was gone.

When the forest around Ed's house died, the neighbors decided to cut their woods down to save their homes from fire and falling trees. The wood was ground into chips and sold. Denuded, the area looked like a perpetual construction site—"Beirut," as the outboard motor repairman put it. Out in the open, visible to their neighbors, the odd little houses weren't so charming anymore, and a cold wind constantly blew in off the bay. Ed and his wife, Sara, moved into town.

Beryl Myhill and her husband, Howard, were part of the post–World War II migration to Alaska. They bought eleven acres above Kachemak Bay in 1946, when Homer was a tiny fishing village without a road connection, and built a trim little cottage among the grasses and wildflowers beyond the treeline. Over the years, a forest of big, sturdy spruce trees grew up around the house. Beryl's favorite was one in which Howard rigged a radio antenna. Her sons built a treehouse in another. The spruce bark beetle stripped the property bare. Beryl's five sons were gone and Howard had died in 1997, so she alone saw their favorite trees hauled off to the chipper and the land open up again into high grass, as it had been fifty years earlier, before their life there together began. Beryl had outlived her forest. She knew she wouldn't live to see it grow again.

Every few weeks, ships carried wood chips from Homer to Japan to make paper, but even the big freighters' cargo holds took only a tiny bite of the dead trees, which covered an area 10 percent larger than Connecticut, mostly on remote public land. The beetle-killed trees remaining didn't stand long. After a few years a big wind would bring them down. They lay piled like pickup sticks on the mountainsides around Kachemak Bay, creating a gray no-man's-land where walking was impossible. So it would remain for a few generations: Ed's inspection of old beetle kill land showed it could become good wildlife habitat in several decades, but the crossed tree trunks rotting on the ground still made hiking difficult fifty years after the beetles had flown.

Spruce bark beetles switched on when spring got warm enough in Southcentral Alaska, and in a few years they may have transformed an ecosystem. There are many such switches in ecological systems, most of them invisible to people used to living within an established band of climate conditions. Newly aggressive pests arose all over Alaska, bugs formerly limited by temperature, some with previous ecological roles so small no one had ever noticed them before. Almost every Alaskan tree species was affected, some over enormous areas of forest. Fires burned more

Alaskan forest, too, with years of two million acres scorched no longer exceptional. Excessive past firefighting was not the cause, as in other western states; there never was much fire suppression in these big Alaska woods. Forests of miserable black spruce, which grow close together atop frozen ground and reach only about twenty feet high in a hundred years, welcomed fire and needed it to regenerate. But increasing the fire frequency meant a faster succession to aspen and many more acres of blackened ground, which, stripped of insulating mosses and shading trees, thawed. Long-stored carbon pumped out of the boreal forest floor. Ecologists theorized this source produced half the CO_2 signal picked up by Dan Endres at the CMDL lab in Barrow, but measurements were just beginning.

Scientists were scarce on these lands below 66 degrees latitude—the Arctic Circle—beyond the primary scope of interest of the NSF's Office of Polar Programs. Ed Berg did his research as part of his job at the Kenai National Wildlife Refuge. If he hadn't set out to find the cause of the spruce bark beetle kill, no one would have done it. But many ordinary people saw what was happening. Beryl Myhill remembered heavy snowstorms that used to blanket her land in Homer a few times a year. Now it just rained all winter. In Anchorage, to the north, winters were getting more like Homer winters used to be—spotty and inconsistent. Farther north, in Fairbanks, an Anchorage snow plow operator showed friends how to move wet snow, the kind that Anchorage commonly received but that was a relatively new phenomenon in the dry, cold Interior.

Alaska Natives all over the state noticed changes. As far as I could learn, no climate scientists were working with them except in the Arctic. Alaska Natives other than the Iñupiat didn't have the oil money or political power to get themselves heard. But even without much interaction with the scientific culture, Natives gathered their own observations, at meetings and even in a database: drying lakes, more trees growing on drier land, plant species that had disappeared, less predictable weather, warmer and longer summers, drought, low rivers, failed salmon runs and disappearing fish, weird new sea life from southern waters, berries ripening too fast and spoiling, warm winters, inadequate snow, dangerously thin ice on rivers, bears not bothering to hibernate, lakes failing to freeze solid, leading to increased beaver numbers and more dams that damaged fish populations. And more. George Yaska, of the Tanana Chiefs Conference, spent his days hearing these stories in villages up and down the Interior rivers. He said, "Just living out there all the time, you can't miss these changes."

Interior forest research happened largely at the Bonanza Creek Long Term Ecological Reserve, a sister site to Toolik but entirely different. Bonanza Creek was merely a plot of land on a forest ridgetop off the highway south of Fairbanks, its only permanent facility a steep dirt road. A visiting dignitary from Washington, D.C., sixtyish, took a tour there in midsummer wearing a pastel silk suit and high heels. University of Alaska Fairbanks ecologist Glenn Juday drove one of the university's four-wheel-drive pickup trucks and I sat behind the two of them. Rain had lubricated the road with mud and Glenn struggled to keep the pickup from sliding into the ditch while delivering an instructive monologue about his work. The VIP wasn't taking much in, however, as she pushed herself back into her seat in fear. We both got out with relief at a dirt lot at the end of the road and followed Glenn down a short path to look over a sweeping view below the ridge, a view like a map where he could point out the forest patterns he was trying to explain. But the VIP couldn't concentrate because she was wiggling and waving to escape mosquitoes that had moved in with an incredible vengeance. She gave me a look behind Glenn's back and said, "Impressive," with a wry smile.

After dropping off the visitor at the president's office, Glenn had to wash the truck to avoid the wrath of the university's fleet mechanics. I plugged twenty dollars in quarters into a vending device at a drive-in car wash while Glenn operated the spray wand and politely answered questions from a German tourist evidently amazed by the process. Then I followed Glenn down a maze of cinderblock corridors in an outdated building to his narrow, crowded office. The posh, silvery IARC building was right across the street, but in the lower scientific caste of forestry, professors made do with what they were given.

Glenn had published important papers based on tree rings, with findings about paleoclimate, his own climate cycle (as yet unnamed), and the fate and future of Alaska's forests. He was a favorite of journalists, receiving enough national coverage to generate real resentment among his colleagues, because his work covered issues that people cared about and because he spoke articulately, plainly, and on many levels, unafraid to bring in the broadest view and philosophical speculations. Consequently, he was often misquoted—there wasn't an up side to doing interviews—but he seemed to enjoy talking. He used his voice well and cracked a big smile when he made a good point. I enjoyed listening to him, so we spent a lot of time that way. When I reread my notes, I realized various ideas I thought were mine actually came from those talks.

Mainly using analysis of tree rings—by physical measuring, density testing with X-rays, and taking the balance of carbon isotopes—Glenn and his primary collaborator, Val Barber, had found systems in spruce trees that allowed forests to synchronize growth and reproduction with cyclic climate changes. For example, white spruce saved energy one year to make seeds the next, when hot, dry conditions would produce fires to prepare the ground for planting. Black spruce growth could be predicted, they found, by the temperatures in April and in the February of the previous year. Hybrids mixed these calendars.

Climate change had thrown off this synchrony and Interior spruce trees were highly stressed. Glenn believed white spruce were at the edge of their viability. And then what? By simply moving lines of climate and ecology, you could end up with grassy steppes or aspen parklands in central Alaska. Glenn thought about such futures but would not predict them with certainty. Science had not witnessed a climate-driven turnover of an ecosystem before. A system probably could disintegrate quickly—forests could die and burn and fail to regenerate, widening patches of change—but it would likely take a long time for a new system to reach equilibrium in its place.

These changes might be happening first and fastest in the north, but they had not been confined to northern latitudes. A persuasive article in *Nature* early in 2003 applied statistical analyses to published studies covering more than 1,700 species globally, using the theory that combining many findings should wash out the biases and confounding influences that turn up in individual research projects. The authors, Camille Parmesan and Gary Yohe, found species ranges shifted 6.1 kilometers toward the poles per decade, and spring events came 2.3 days earlier per decade, with a statistical level of confidence above 95 percent.

Of course, the spruce bark beetle proved that climate change doesn't march ecosystems steadily across the landscape; it can also cause them to convulse and suddenly transform. But assuming a general trend toward the poles, how would ecosystem migration look? An ecologist in Alaska might see a new natural world created outside the window. Without human barriers to animal and plant migration, living systems would be free to rearrange themselves across the land as dictated by the new patterns of climate. World agriculture could probably adapt, too, Glenn said, although perhaps not individual farmers. Maintaining artificial biological systems such as wheat fields would essentially mean sending the seeds to a new address. But the future was troublesome for natural systems in the developed world's temperate latitudes. When ecosystems disintegrated on biological islands such as the national parks and national forests or on any undeveloped land surrounded by the barriers of civilization, the genetic material needed to grow a new natural system might not be readily at hand.

"Climate always changes, and it will change, and we've got nature at fixed addresses," Glenn said. "It's up to us. We'll have to move the organisms. Basically, we'll become ecosystem doctors. We'll be doing ecological interventions at an order and to an extent that we can't even imagine now."

Glenn imagined a modern-day Noah, and he was willing to talk about it—just the sort of behavior that made some other scientists uncomfortable with him. But he couldn't help it. Glenn's idea of professorship bound him to an inviolate oath to tell the truth, so he testified. And his life experience had taught him a lot about land conservation. The son of an Indiana hog buyer, Glenn grew up with a love of the woods that took him into forestry in the late 1960s. In graduate school in Oregon, he helped create that state's law on protecting natural areas. He came to Alaska in 1977 to join the work on the biggest conservation law of all time, the 1980 Alaska National Interest Lands Conservation Act, which set aside an area as large as California and closed the Alaskan frontier. As his small part of that immense decision, Glenn, still a young man, made quick judgments concerning enormous tracts of land, enacting on the fly his ideas about habitat fragmentation. The Alaska land use patterns that emerged could hardly have been better prescribed for adaptation to climate change, with big, connected preserves for ecosystems to move without barriers.

His current work untangling the processes of forest ecosystems meant no less to Glenn than reaching toward God, both in revealing His mysteries and in helping solve the great problem He had allowed to challenge humankind.

Glenn said, "We've described a must-solve problem that can only be solved by greater solidarity, cooperation, advancement of knowledge, common action, unity. So it will require humanity to organize itself, to communicate effectively, to negotiate and trade and find spheres of mutual advantage, to work through these problems—when our own natural inclinations would not have been to do that, right? We would have gone the other way.

"In that sense, it fits in perfectly with my worldview. God is active in history right now."

Glenn invited me to dinner at his home in a tidy 1970s subdivision near the Chena River. He and his wife, Mary Beth, had four children, three of whom had left home already and one who never would. Collin was sixteen, with whiskers and hairy arms, a gentle boy with a bemused smile and the mind of a three-year-old. In the area of his father's greatest strength he had a complete deficit—no imagination at all. This evening after work, as each day, Glenn sat on the living room carpet in his short-sleeved button-down shirt and pressed pants, with rosary beads hanging out of one pocket, and put together Brio wooden train tracks. Collin couldn't build them himself but enjoyed driving the little wooden trains around on the routes Glenn made, naming the streets and intersections of Fairbanks as he went.

Mary Beth ran the bookstore at church, doing only $10,000 a year of business but gaining the opportunity to stay up with the latest in Catholic theology. The couple had met in college, where they played folk music in a church group. They had given up that style of worship only a few years back when they began attending Latin mass instead. Glenn's Catholicism was rooted as firmly as a windswept treeline shrub. As a child in a Protestant community, he had endured beatings and daily taunting. That trauma seemed to disturb him still and to stiffen his back as a member of an intellectual minority in the scientific culture.

At dinner, Glenn and I discussed the unknowns in the succession of black spruce, birch, and aspen and the spiritual history of the Athabaskan Indians. Collin interjected with "Saturday, July 13," and "Six plates," and "Partly cloudy." He had an eye for dates, numbers, and weather and enjoyed repeating facts. When finished with dinner, he set up a line of buckets along the eaves of the house. He had been prohibited from playing with the hose because he made too much of a mess, but he was allowed to play with any water he caught in buckets. Tonight, however, rain looked quite unlikely. Glenn hoped Collin might someday be able to do a simple job related to weather. Mary Beth said, "We know Collin will always be with us."

At our last one-on-one talk in his office, Glenn stated, with as little relation to the rest of the conversation as any of Collin's remarks that night at dinner, that what makes humankind unique is not our intelligence but our ethical capacity and the knowledge that we were created out of the love of God. I asked him to explain how he reconciled science and faith. This wasn't as much fun as our earlier talks. Glenn was as open and articulate as ever but seemed nervous about what would come of it. After explaining why the commandment to keep the Sabbath was more important than the one not to murder, he said, "I suppose you would consider me a fanatic, but I'd like to think of it as just a coherent worldview. You may think it's wrong. Fine. That's all right. It won't hurt my feelings. But it does cohere."

Glenn saw no division between religious faith and rational science; in fact, he said they were inextricable. He pointed me to the writings of Catholic philosopher Stanley Jaki, who reasoned in the 1970s that it was Christians' linear, beginning-to-end view of the universe that gave form to Western science on "the single intellectual avenue forming both the road of science and the ways to God." This scientific road, Jaki wrote, started not in the Renaissance but in the Middle Ages, when friars used rational thought and study for insight into an intelligible, rational God who had made everything. Other cultures with fatalistic and cyclic worldviews failed to spawn ever-advancing scientific achievement because their spiritual beliefs were incorrect. Jaki cited "the logic-defying aphorisms of Taoism" as the height of this antirational error: "Today the same insults to the intellect are the daily fare of Chinese minds desperately trying to live with the consistency of science and with the inconsistencies of Taoism resuscitated in the form of Marxist-Maoist dialectic." (This was published in 1978, before Marxism and Maoism became history.)

The invigorating thing about these ideas was that, like science, they made definite, testable statements. So what about Galileo, who in 1633 was forced by the Inquisition to retract his correct scientific findings? (The Roman Catholic Church formally accepted the validity of Galileo's work in 1993.) Glenn said the Galileo story taught in school was a myth; Jaki pointed out that Copernicus, Kepler, and Galileo were all good Christians who needed their faith to make their scientific advances. The fourteenth-century writings of William of Ockham posed a more fundamental challenge. Ockham, excommunicated for heresy, wrote that human reasoning should start with our direct perception of reality. The handy mental tool of Ockham's razor, still influential in science, holds that one should not call upon hypothetical entities to explain empirical evidence unless there is no alternative; in other words, choose the most direct explanation that works. This thinking led Ockham to deny that God's existence could be proven by human reasoning and helped make him popular two centuries later with Martin Luther, two reasons Stanley Jaki couldn't stand him. Jaki's argument against Ockham essentially charged him with reductionism, the error of breaking observations and reasoning into parts so small they lose their meaning.

But much of the success of modern science grew from breaking down nature into its tiniest parts. In some fields, including biology after the sequencing of the human genome, science seemed to have gone as far as possible in that direction, and now a new movement of search and synthesis was dawning. In biology and ecology, as well as in climate change research, computer modelers were attempting to rebuild systems out of the parts that science had previously dismantled. The invisible magic of chaos and complexity in a computer's mathematical calculations took the place of a universal hand guiding the world. These computing systems obeyed Ockham's razor and, at least in theory, they yielded broader meaning—albeit meaning whose details were too numerous for a human mind to absorb.

But Glenn said models could produce illusions, too. Like every other scientist I met, he believed models were useful: they were good for testing concepts about systems and they forced discipline by requiring researchers to define their ideas mathematically. But he said modelers could at times act like sophisticated television viewers, so well trained in the analysis of TV commercials' forms and subtleties that they started to find meaning where really all that existed was an internally self-referencing system of flashing images. Practically, models could not predict the fate of ecological systems for the same reason no one could have predicted the unique behavior of the spruce bark beetle in the big kill. Nature contains trip-wire thresholds for sudden change that are invisible until they are crossed. A pest that hasn't been named yet might be the next one to rip across the landscape at a new step up in temperature. In unprecedented conditions, there is no way to say what might happen. Glenn said ecological systems contain too many species acting in too many combinations over too much time for a computer model to make predictions, although the striving, rationally driven culture of Judeo-Christian science makes the attempt irresistible.

Like other scientists, Glenn fell back on intuition, the grand human calculator still able to challenge computers at chess and easily able to surpass them in analysis of unfamiliar real-world problems. Such as hog buying. Glenn told how his father would spend his day at the stockyard, knowing each meatpacker's preferred style, quality, and quantity of meat, looking at hogs brought in by many farmers, evaluating their characteristics and numbers, then consulting with partners at other yards on the bidding they were seeing. "At the end of the day, he'd take one last walk down the stockyard and he'd make his final deals," Glenn said. "Think of all the judgment factors that go into that. It just boggles the mind. And that's what it does to ecological models. It just gets too many factors and it breaks down."

He said, "Are there things that it is not possible for the human mind to know? That is part of revelation. We just can't get our minds around God as an infinite, all-powerful, all-knowledgeable being. And [in ecology] it's an open debate. I tend to side with those who say ecosystems are not only more complex than we know, they're more complex than we *can* know."

A great hunter and his wife lived alone, unaware of any other people on earth, lonely and without diversions. They had a son and raised him to be a skilled hunter, but one day while hunting alone in the mountains, he disappeared and never returned. A second son met the same fate. Their third son also became a strong hunter, and while he was hunting caribou one day, a large young eagle circled and landed and, pushing off its hood, became a man. The eagle explained that he had killed the two brothers because they had refused a gift from the eagle mother of song and feast, which he now offered to the third son. This young hunter didn't know the meaning of "song" or "feast," but he agreed to learn, so they journeyed together over the flat land, the river canyons, and up into the mountains, where they heard the mother eagle's heart beating as loudly as a crashing hammer. She was old, sad, and decrepit but brightened upon hearing that the young hunter was willing to learn how to build a qargi, to sing, to perform the ceremonies of great present-giving feasts, and to make drums that would beat like the eagle mother's heart.

"But we don't know anybody but ourselves," said the young caribou hunter.

"Human beings are lonely because they have not the gift of festivity," the eagle mother said. "Make ready as I have told you, and when everything is prepared, you must go out and look for people, and you will meet them two by two; but you must get them together until there are many, and invite them to the qargi; then you must hold a song-feast."

Back home, the young man and his father worked hard to build a large qargi and hunted to gather a great hoard of food and gifts. Then the young man went out in search of people and found them as predicted, two by two, all dressed in furs. They feasted richly and received the hunters' gifts, then danced, sang, and joked all night, their voices joining in a great, joyful chorus. Only as day broke did the guests depart, each falling on all fours while leaving the qargi and returning to their real shapes as animals to go back into the world.

Later, the young eagle again met the hunter and asked him to return to the mountains. Climbing to the mountaintop, the hunter met the old mother eagle. She was no longer bent and worn. The joyful celebration had made her young again.

The Kivgiq, or Messenger Feast, helped keep the eagle mother young when villages of Iñupiat gathered every few years for festivals of song and dance, jokes and stories, rites performed by shamans, and vital trade. It reenacted the story of communication between the species and was the Iñupiat's most important social event. Two runners, the messengers, were sent to a related village with an invitation to Kivgiq, generally held in late winter, around the time the sun reappeared from its two-month absence. The messengers carried staffs marked with rings that stood for invitations to individual guests and requests for certain gifts. The hosts would reciprocate with gifts of equal value, or even bid up the value by giving more. This was one way the inland and coastal villages carried on commerce in the goods they needed to exchange. But despite its spiritual and economic importance, Kivgiq was fun and lighthearted, with practical jokes, races, and impromptu humor, special songs and comical traditions, and gift giving and excess that went on for days or even weeks. Villages and individuals composed their own songs and dances for each Kivgiq and spent weeks practicing—creativity was part of their spirituality.

Traditional Iñupiat would not have called Kivgiq a religious meeting, because their language had no word for religion (modern, Christian Iñupiat would avoid calling it that for other reasons). Traditional Iñupiat did not distinguish between spiritual and practical life. They performed rites and observed rules every day to negotiate a world in which the animals they subsisted on had spirits similar to their own—complicated rules that limited the kill of certain animals, helped hunters demonstrate respect for their prey, and often required a lot of effort and sacrifice. Their shamans communicated with the spirits of animals and people and traveled to the moon for help and insight. In a world without coercive authority but full of magic, the Iñupiat did not conceive of a supreme being, but they were careful about maintaining their social relationships in the spirit world that underlay and controlled the physical world. The qargi was a church and a social hall, and the two functions were one.

Iñupiaq creation stories tell of human heroes actively working to make a world, sometimes, as in the story of Kivgiq, with the help of animals. In one story, when the world was flooded, a green patch of land intermittently appeared above the waves but always disappeared quickly. A young whaler kept his crew awake until the chance arrived to spear the moving piece of land, which then rose up and became the broad expanse of tundra. The sea worms escaping from it became the rivers. People began to age. In contrast, humankind in Genesis is passive. In that story, God made man in his own image, with an immortal soul, gave him animals and land for his use, and set him up in a paradise where even childbirth was painless. Eve's decision to transgress by picking the apple was her species' first independent act. The price paid was everything bad that happened ever after. The two stories neatly summarize the two cultures' concepts of nature and of authority.

In light of such fundamental differences, the task Presbyterian minister Sheldon Jackson set for himself in the 1880s—to erase all Alaska Native spiritual beliefs and replace them with Protestantism—would seem madly ambitious. But Jackson had reasons, both spiritual and worldly. Elsewhere, American Indians had been segregated into reservations as inferiors, but Jackson believed Alaska Natives were a better sort who could eventually become equals of whites, a progressive idea for the time. Christian theology would help establish Christian civilization, wiping out the incompatible Native ways sanctioned by their free, egalitarian society. Missionaries were especially disturbed by guiltless Iñupiaq sexuality and nudity. Imposing monotheistic discipline and hierarchy would help mold the Iñupiat into hourly workers and nuclear families in the American style. Jackson expected Natives would ultimately switch to American cultural norms in language, religion, dress, housing, work habits, marriage, and livelihood and become worthy of full citizenship.

In 1880, he convened a meeting of leaders of Protestant denominations in a New York hotel room to divide up the work, assigning regions of Alaska to each church (but not to the Russian Orthodox, who for a century had taught Christianity in Alaska Natives' own languages without demanding cultural conversion as well). Benjamin Harrison, a U.S. senator and later president, was Jackson's close friend and saw his ideas enacted into law and Jackson himself appointed czar of Alaskan education in 1885. The missionaries who fanned out across Alaska had the force and funding of the federal government behind Jackson's plan of religious, cultural, and economic conversion. They used it to settle mobile people into fixed villages, to consolidate villages (Nuvuk was abandoned for Barrow), to do away with warm sod houses in favor of cold, drafty frame houses, to suppress Native languages, and to ban traditional stories, songs, and dance. Lela Kiana Oman, an elder originally from Noorvik, told me how her father would gather his children around him late at night to secretly pass on ancient stories that had been outlawed, charging them to tell no one. She lived long enough to carry those stories through that dark period and to write them down in her old age.

But it didn't take physical coercion to introduce Christianity to the Eskimos; at a certain point the new faith took off like wildfire. Religious rules and prayers introduced in the Moravian zone on Kotzebue Sound spread from mouth to mouth up the Kobuk and Noatak rivers to arrive, somewhat distorted, in the central Alaska Arctic. Vilhjálmur Stefánsson, traveling in 1908, reported the Eskimos converted in a matter of a year or two, but the Christianity they at first adopted was really just an extension of their previous view of nature. In the traditional Iñupiaq spiritual view, failure to observe the proper rites of respect for animals could mean the next hunt would fail or even be fatal. Rules and taboos multiplied when they seemed efficacious. Whalers abandoned their talismans when they saw white whalers succeed without them—they had always been a means to an end. The prayers and taboos brought by the missionaries at first simply added to this system: one didn't hunt on the Sabbath, one wore clothing even indoors, one prayed for caribou, all to ward away bad events and to encourage safety and success. After a missionary in Kotzebue Sound chided men for fishing with nets on the Sabbath, word passed across country to the Colville River, far to the east, that God forbade nets on Sunday, so on that day they fished with hook and line instead.

The Iñupiat had lived with spiritual rules as solid facts and viewed Christian teachings the same way, not as guidance or pretty ideas but as unbreakable requirements. Stefánsson described a man who left family members in the cold to freeze so he could make it to the village before the Sabbath. The Iñupiat in the village held back from rescuing the stranded party until midnight passed on Sunday night. Their relatively liberal Presbyterian missionary tried to explain that the Sabbath did not have to be kept so literally. For such laxness, the village elders wrote to the Board of Home Missions in New York and asked to have him removed.

It may be that no explanation is necessary for why the Iñupiat converted to Christianity so quickly, as the faith often spread rapidly among indigenous peoples. One simple reason for this, pointed out by Glenn Juday, was that the capricious spirits of the naturalistic universe offered little solace in the face of death, which came so brutally and unexpectedly, while missionaries offered eternal life in exchange for accepting a loving and forgiving God. The more skeptical Stefánsson reported that the Eskimos appended heaven and hell as new wings of the universe they already knew, applying the magical mechanisms of Christian prayers and observances as a reformulated style of shamanism to attain immortality and avoid damnation. Charles Brower offered a yet more practical explanation, also related to the fear of death: he said the Iñupiat in Barrow converted when they saw the missionaries' medicine worked better than their own shamans' healing. Sheldon Jackson may have had all three of these forms of persuasion in mind when he set out to reform Alaska Native society in a mainstream American mold. The coercive power of death itself may have proved stronger than even his federal backing.

What the Iñupiat believed about death before the missionaries came is mostly lost. If the elders ever divulged the information to outsiders, it was never recorded on paper, except for a vague idea that dead relatives would return. Students of Iñupiaq tradition such as Jana Harcharek could turn only to Dr. Simpson's 1855 monograph, based on his interviews aboard the *Plover* with Erk-sing-era, whose few contradictory lines on the subject are all that survive.

People in Barrow eventually became devout and sophisticated Christians. The topic of how that happened was not a polite one to bring up. In the eyes of some Christians, animal spirits—and certainly shamanism—were satanic. When I asked Savik and Myrna Ahmaogak why the Iñupiat converted so quickly, they looked at me silently, and then Myrna said, "The Holy Spirit." Arlene Glenn politely refused to discuss the subject of Iñupiaq tradition and Christianity. Only with Jana Harcharek did I get to the point of asking how Christian Iñupiat reconciled their traditional worldview with the opposite view in Genesis. She was visibly uncomfortable as she explained Kivgiq and how so much was lost when missionaries labeled ancient teachings as spiritually bad, and yet how they persist.

"On the other hand, some of the Christian values fit in very well with the values of Iñupiaq people," she said, repeating a stock phrase, somewhat stiffly, that I had heard several times before. And then she relaxed and said, almost imploringly, "For me to talk like this, it takes a lot of courage. Because it's still very, very difficult for our people to accept their traditional religion." And then the interview ended.

Almost a century after Barrow converted to Christianity, the Iñupiat still believed that animals had unique spirits like human beings—not as a rationally derived ethical position but as an actual, tangible fact. This belief made catch-and-release fishing grossly offensive in Iñupiaq eyes. If an animal wandered to an Eskimo's door, it would be wrong to reject the gift: the animal must be killed for food. Hunting laws intended to uphold concepts of sport and fair chase were wrong; killing fellow animals was not a game, and it should be done as simply as possible. Likewise, they had little respect for adventurers seeking hazards outdoors as a form of recreation or for environmentalists who viewed the wilderness as a realm of spiritual purity that man could only sully. What others saw as a playground or a church they saw as home. Iñupiaq hunters still felt kinship and real admiration for polar bears, foxes, wolves, eagles, and other predators with capabilities they lacked, like colleagues admiring a co-worker facing the same challenges.

People who had survived on a diet purely of animal flesh believed the animals they killed shared their own claim on eternity, and that was one of the reasons they survived. Richard Glenn said respect for nature and its inhabitants—and respect's partner, humility—kept people alive. Down south you might have been able to pretend you weren't cold; here your life depended on admitting you couldn't tough it out. Trying to get home from camp in time for a meeting could kill you, too. The oldest story in the annals of Alaska wilderness tragedy was of people pushing their limits against bad weather because they had to get somewhere by a certain time. The elders even counseled against running up and down the beach chased by the waves: it was disrespectful to the sea.

"The biggest connection between traditional knowledge and the spiritual way of life is about respect; respecting the environment, respecting the land, respecting the animals," Richard said. "Traditional knowledge to me is centuries of trial and error. So what looks like an elegant solution is something that has only been learned because we've tried to do it in the wrong way in the past and this way works better. And that is also built around respect. Safety is built around respect. Survival is built around respect. You think you're better than the weather? Let's see what the weather has got in store for you."

Now the weather was changing for everyone, and there was a sense, in what some elders said, that a lack of respect was indeed the reason. This had been predicted. Catherine Attla of Huslia, an Interior Athabaskan village, said it this way at a meeting in 1999:

> We were put on this earth to have our own way, our own place. My stepfather Bobby had a radio. We were in the house—quiet—the radio was on and we didn't understand one word. Pretty soon we hear that crane in the middle of winter. That's not right. You never make a noise of an animal in the wrong season. You don't talk about how robins make noise in the winter. He died in 1959, but grandpa chief Henry took his place. I started to hear him tell stories. He was living when they first walked on the moon. He told people, "That's too bad that people have to do that. You young people, the weather's not going to be at the right place at the right time."

CHAPTER TEN

The Challenge

IN THE FALLTIME AT POINT BARROW the sea was free of ice and waves rumbled ashore on fine gravel. Sometime in October or November icebergs would drift to the beach and the calm, green water in between would freeze and set them in place, a green floor between white and blue sculptures. But in late September, with just a dusting of snow on the tundra, the ocean was a rich blue-green under a clear, deep blue sky. The weather was so good Savik Ahmaogak didn't want to come back from camp upriver. The family discussed staying there at their newly expanded cabin, fishing and hunting. But Savik said no, they had to whale, if only for Point Hope. That large, traditional village had caught no whales at all in the spring season and in Point Hope, unlike Barrow, there was no fall season to make up for it because the fall bowhead migration didn't concentrate close enough to shore for hunting. Barrow whalers wanted a good season in part so they could charter a big cargo plane to send maktak and meat to Point Hope and the other hungry villages.

Whalers from Utqiagvik (later called Barrow) pursued bowhead in the fall in old times, but gave it up before Lieutenant Ray's expedition arrived in 1881. Fall whaling was always harder because there was no ice from which to ambush passing whales; crews had to patrol the big ocean in the umiaq, hoping to get lucky among relatively dispersed pods of passing whales depleted by whaling ships that also hunted in the fall. Besides, with whaling poor, it was more profitable to trade with the Yankee whalers. Roy Ahmaogak told me proudly that Savik Crew was among the first crews that helped to revive fall whaling in the 1980s. They launched their aluminum boat, and later larger boats with bigger motors, to patrol autumn waters through the twelve hours of daylight, looking for a spout and counting on horsepower to get them to the whale before it could escape. Today the fall season was often as big or bigger than the spring. Whatever quota whalers didn't fill in the spring they would take in the fall. That year, with the spring season nearly a bust, plenty of quota was left for Barrow.

For me, returning to Barrow that fall felt like a homecoming. It was beautiful: the tundra, marbled green and white; the lakes, mysterious blue-black-green, reflecting the sky; and the sea, so large. Back at Myrna's table, with the family photographs and documents under the glass and the food always passing, Savik Crew greeted me like a brother. Everyone gathered to look at my photos from the spring and marked the backs with their initials for copies. I settled in at the ARF, where a young biologist was devouring paperback novels while waiting for whaling to begin.

Craig George arrived at his office at NARL on a four-wheeler all-terrain vehicle looking like an old-fashioned aviator in his goggles and flap hat, wearing big mittens and bulbous Arctic bunny boots. The International Whaling Commission was in the process of quietly reversing its decision to ban Iñupiat whaling, but Craig remained bitter about the meeting in Japan. After twenty-five years in the Arctic, he had some doubts about his life. Barrow was a hard place to get older; if you weren't outdoors, you were sedentary, and without the toughness of youth it was hard to get outdoors. Over the summer, Craig had taken his wife and two boys to visit his famous and vigorous extended family in the Tetons, where they had gone hiking and camping and gathered at an outdoor elk roast, and he had found himself thinking, "God, what am I doing to myself?" He said, "I don't ever think about it until I leave [Barrow], and then I think about what I'm giving up.

"I've been thinking about that a lot. My kids have no understanding of their culture, as far as I can see. They think they're Native. They really do. As a matter of fact, we kind of argued with them about it when they were younger. They insisted they were Eskimo. They absolutely don't see color. We had to tell them about racial differences and stuff. But I would like them to know something about their roots. I'd like to get them out for a couple of years, somewhere, just so they know it. Because they're not going to know it if we stay a whole lot longer.

"There's a lot you don't have here. The main part of my culture I miss is being able to go to one of those western bars and hear some live music. Have a beer and listen to a folksinger. Of course we have to do it ourselves. We play a lot of music, but it's at someone's house. But just having roads, and being able to go outside without dressing up all the time. That Wyoming weather is just astonishing. No mosquitoes, no rain, nice temperature, hard, flat ground. No mushy, gushy, muddy stuff. We went camping in Yellowstone, and thought, 'God, what's wrong with this? There's no wind, no bugs, the ground is hard and flat and dry, there's firewood everywhere.' We felt like we were in a studio somewhere."

A plague of polar bears had landed at Barrow, and Craig shared the dicey problem of figuring out what to do about them. The biologists didn't want to kill the bears but preferred not to tranquilize them either, as it rendered the meat inedible if the bear was killed later, a terrible waste in the Iñupiat's eyes. One recalcitrant 1,100-pound bear in the way of children going to school was shot, and there were another sixty or seventy around.

Over the years, the North Slope Borough had invested considerable money and effort in protecting people from polar bears. Even in tiny villages of a few dozen houses, school buses carried children to keep them safe from bears. In Barrow, the city buses were tracked by GPS and their positions constantly broadcast on a map on cable television so riders wouldn't have to wait outdoors with the bears. Villagers learned to leave hunting waste at designated points outside of town where bears could scavenge without encountering people. In Barrow, falltime whalers deposited bones and viscera at Nuvuk, the tip of Point Barrow, where polar bears could reliably be found feasting when the sea ice, their normal habitat, came near enough for them to disembark.

Not many people had been killed by polar bears, but plenty had been chased. A man was roughly normal-sized prey for a polar bear, and the bears' reasons for abstaining from human meat were unknown. Other game may simply have been easier.

In the village of Point Lay, in December 1990, when it was dark all the time, windy, and –35 degrees F, Carl Stalker and his eight-months-pregnant wife, Rhoda Long, left a family member's house for the short walk home and encountered a hungry young male bear. Rhoda screamed.

Carl told her to run while he pulled out his pocket knife and ran the other way to lead the bear away from her. An hour later, villagers found a bloody spot in the snow where the bear must have caught Carl near his doorway. Carl himself was gone. Fearfully, with hesitation and much discussion, they formed a tracking party to pursue the bear in the dark, following a trail of blood. The bear was found in the beam of a snowmachine's headlight, out on the sea ice, hunched over Carl's bones, which in two hours' time had been stripped of flesh. The party shot the bear and found the ineffective marks of Carl's knife on its eight-foot body. "It was just a little eight-inch pocket knife," said his brother, Jacob Stalker, Jr. "He wasn't going to let that bear kill his wife, with the baby inside her."

Taqulik Hepa, the Wildlife deputy director, told me how her parents, Tom and Margaret Opie, arrived at their cabin upriver one July and were charged by a polar bear that chased their boat until they killed it. In that situation, as currently in Barrow, the problem was a stranded bear. Polar bears belonged on the ice pack, where they could hunt seals. On shore there was little for them to eat and hunger could make them dangerous. Near the southern edge of the bears' range, in Churchill, Manitoba, on Hudson Bay, bears commonly came ashore during an ice-free summer, fasting until fall. Warming temperatures brought spring progressively earlier and researchers found the bears growing skinnier and less healthy in clear synchrony with the earlier loss of ice.

In Barrow in September 2002, the ice neared shore and bears got off, perhaps congregating in anticipation of the impending whaling season that would leave fresh food at the bone pile. Then the ice quickly withdrew. Bears can swim a long way, but not far enough to reach an ice pack hundreds of miles to the north. Tourists and locals went out to the point in all-terrain vehicles to take pictures of dozens of bears feeding on the bone pile. Bears picked over year-old leavings on a mound the size of a building. When the bears started coming into town, authorities shut down bear viewing on the theory that off-road vehicles might be chasing them toward Barrow. That made tour operators mad. Craig George and the other biologists were in the middle of it and losing sleep as they dealt with bear sightings. An eccentric known for walking around town had disappeared, and people wondered if he had been eaten.

The sea ice was far, far away. Only a satellite image could show exactly how far, but the stranded bears and the big waves on the beach told the story generally. The growth of waves on the ocean follows a simple formula with three variables: the strength of the wind, its duration, and the distance across which it blows, called the fetch. Even a long windstorm won't build big waves on a pond, because there isn't enough fetch. In former decades, when the sea ice usually stayed relatively close off Alaska's Arctic shore, big waves came infrequently. But now the sea opened wide in the summer and fall and a big sea swell had a chance to build—long, powerful waves that could roar onto the beach. Erosion advanced quickly, as much as 1,500 feet over a few decades on some parts of the coast, but since Alaska was poorly mapped, it was hard to say exactly how much the rate of erosion had increased. Villages that were washing away certainly believed they were victims of climate change. And hunting was more dangerous. A few years earlier, a boat from Nuiqsut was dragged under by these big waves while pulling home a whale.

Some Barrow whalers claimed they had already adapted to climate change: they had bought bigger boats for fall whaling. Knowledgeable white men such as Craig George and Glenn Sheehan foresaw the next step of adaptation as a forced phasing out of spring whaling to hunt only in big boats in the fall. But Richard Glenn cringed when I suggested that prospect to him. He loved camping out on the ice.

Roy Ahmaogak and Richard had built a second floor on Savik Crew's upriver cabin that summer and began digging a new ice cellar for the fish and caribou they caught, photographs of which Richard showed me with great pleasure over civilized little glasses of scotch at his kitchen table while Arlene handled baby Joanne in her rocking chair in the adjoining living room. To get to the cabin, they drove the boat east seventy miles along the coast, outside the barrier islands, then turned to the south ten miles through the intricate delta channels of the Ikpikpuk River, then branched off onto the Mayuagiaq River, tracing its meanders more miles through level green terrain to where the rectangular box of the cabin stood by a loop of smooth water. The land was mostly water there, near twenty-five-mile-long Teshekpuk Lake, with thousands of lakes, streams, and swamps interconnecting over scores of miles in every direction. The joy of that place, familiar to me as a father, lay in the presence of Richard's entire family in a self-contained space, the mixing sounds of their common life, and the successful effort they shared in meeting the requirements of survival. Camping in the wilderness with family, everything was an effort—washing dishes, protecting children from hazards, obtaining fuel and warmth—and from that effort came the pleasure, as generations joined in work as one. That was the smile I recognized on Richard's face, and likely the reason Savik didn't want to come back to town after an entire season upriver.

But Richard's talents so burdened him that he never managed a long, satisfying stay at the cabin over the course of the summer. While others watched the passage of months there, he journeyed back and forth, carrying groceries upriver and then returning to Barrow for his duties in the scientific, corporate, and political worlds. His executive rank was not yet high enough to allow him to be unavailable when called by senators or the mayor. Uncle George Ahmaogak was up for reelection as borough mayor in October, challenged by his perennial foe, Ben Nageak. Some had pushed Richard himself to run against George, but he supported his uncle because of George's powerful voice at a time when the Iñupiat perceived themselves as under attack from the Republican legislature in Juneau. In the midst of a chaotic campaign, Richard ended up working as campaign manager, a demanding chore with a lot riding on it. The borough was larger than many nations, its oil-funded local government was the region's main source of economic life, and the mayor's patronage represented one of the primary doorways to upper-middle-class economic status. The campaign worked through complex family networks, the allegiances of village elders, and the inner workings of groups such as Barrow's close-knit Filipino community, and it required fund-raising and long flights by the candidate and by carefully chosen family members and associates to tiny villages for the distribution of goodies printed with the mayor's whale logo, such as the huge blue coffee mugs.

Some wanted Richard to end up as borough mayor. Some wanted him to end up as the head of ASRC. On the verge of forty, Richard wasn't sure where he wanted to end up. With John Kelley, he was lobbying to get Kenny Toovak an honorary doctorate from the University of Alaska Fairbanks, but the fieldwork on his own dissertation had long ago gone stale. He wanted to whale and to dig the ice cellar at the cabin upriver, but as the Barrow captains' self-imposed starting date neared for whaling, he was in Washington and Arlington. He got back just in time, and then fog and wind rolled in and all boats stayed ashore. And the next day he had to attend an ASRC board meeting. And before whaling season was over they wanted him as a delegate to the annual Alaska Federation of Natives meeting in Anchorage; no ASRC board member had ever refused to be a delegate. Away too much from the ones he loved, away too much from the activities that sustained him, he tried to stay up with obligations that ran like hungry dogs, without knowing exactly where he was running.

First light on the day of Richard's board meeting showed an ocean perfect for whaling. A rich rim of gold emanating from the eastern horizon held up a sky dome of velvety blue. Roy and his crew climbed the high gunwales of Savik Crew's fall whaling boat while it still sat on the trailer and Richard took the wheel of the pickup truck to back it down the ramp on the beach near NAPA. Then he unhitched and left for work. Lots of boats and crews were launching and floating just offshore. The air was sharp with the fall chill and the excitement of the moment amid the scent of the sea and fuel. Roy opened the bomb box and prepared the weapons, heavy gear thumping on the fiberglass deck, sounds flattened by the surrounding sea. Eben took his place with the harpoon ready on the foredeck. In the aft, the grub box yielded metal thermoses of hot coffee and store-bought cinnamon rolls. Roy pushed the throttle lever forward and we flew out into the Arctic Ocean, directly away from Barrow. A long swell from the north lifted us in slow rhythm.

These waves came from far away, built over hundreds of miles of open water where the pack ice had retreated. A few months later, scientists from Boulder confirmed its unprecedented distance when they announced that the ice had shrunk farther than ever before measured with passive microwave satellites first launched in the 1970s—the result of a warm, stormy summer that broke and melted the pack. Arctic sea ice was smaller than the long-term average by a million square kilometers, or 14 percent, and much of the retreat was on the Alaska and Siberia side. The ice also was thinner and less compact, even at its center. Jim Maslanik, a coauthor of the study, said the ice extent was probably the lowest in fifty years. It was as if a continent were disappearing.

Polar bear, walrus, various seals, birds, fish, and lower organisms lived on the permanent floating ice. When the pack ice shrank, its edge retreated beyond the relatively shallow water near shore. Walrus and other bottom-feeding ice animals were separated from their food supply and polar bear were separated from shore. Beyond the continental shelf, nutrients and living organisms falling from the melting ice edge sank into the abyss instead of nourishing clam beds that thrived nearer the surface. The icebreaker cruises that voyaged to study this system in the summer of 2002 had trouble even finding much solid ice in the area they planned to research. If the ice ever disappears entirely, its ecosystem will likely disappear, too. With world ecosystems migrating toward the poles, the ecosystem nearest to the pole would be pushed right off the planet.

The Northwest Passage between the Atlantic and the Pacific oceans was sought by Sir Martin Frobisher in 1576, took the life of Sir John Franklin and his men in 1845, and was successfully navigated for the first time only in 1906, by Roald Amundsen after a three-year journey in the little sailboat *Gjöa*. But as whaling season began in 2002, a large, steel-hulled sailboat anchored off Barrow had just made the passage without drawing special attention from the Eskimos, who said someone seemed to do it every year. Maritime experts expected that within a decade the Northwest Passage and the Northern Sea Route over Russia would be navigable for a month a year in unreinforced vessels. Military and diplomatic conflict would likely ensue over the unresolved national status of the waters. A passage navigable year-round by tankers and the like also presented the dreary certainty of eventual Arctic oil spills.

Savik Crew's twenty-three-foot boat had high sides and a heavy, V-shaped hull that cut steadily through the waves. A new engine Richard bought after burning up the old outboard on a run to the cabin could drive the boat faster than forty miles per hour (it was a 225-horsepower Yamaha four-stroke). The boat was probably built for sunny fishing trips, Bermuda shorts, and cases of cold Budweiser, but now it looked like an Eskimo whale boat: Eben forward with the harpoon and Roy at the wheel, his round brown face scanning the horizon while the wind whipped the fur trim of his white hunter's parka. Roy liked to hunt his own area, and so we drove west, thirteen miles straight away from land, leaving Barrow over the horizon behind us and entering a huge, empty circle of blue sky and waves. Roy gazed at the distance for spouts. My eyes were untrained for looking long and patiently for something that wasn't there. I scanned, but then I found my mind wandering and my gaze floating hypnotically on the water. Hours passed without anyone's saying a word.

Roy saw something, spun the wheel, and pushed the throttle to full power. With an exhilarating forward surge the boat skimmed over the waves, slapping their crests. In this newer style of whaling, power replaced the patience of lying in wait on the spring ice. Crews could see a whale blow a long way off: with enough speed, the boat could be there when the whale came up for breath, or even before it went down. Once in the area, however, the boat would slow and guile would replace power. Like the great whaler Malik, the captain now needed the skill of following, judging where the whale would surface so he could put the harpooner right on top of it. There must have been whales in the spot where we arrived, because other boats had seen them, too, and gathered around and slowly circled in hopes of a shot. But no whales surfaced.

The waves were rising and some boats were going in. Roy took us north along the shore into larger seas. The boat could operate, but it was hard to see how you could tie onto a whale and safely tow it home in such waves. Rounding Point Barrow we leapt steep, five-foot waves where the sea swell was piling up on the shallows. Looking back toward the point, where the village of Nuvuk once stood, I could see the bone pile and the loitering polar bears. One ran down to the surf as if to guard the point from the breakers frothing ashore.

The next morning the wind and fog were back and many boats stayed home, but Richard had to go out. The Savik Crew's boat was more seaworthy than most. Many whalers used old, riveted aluminum lake boats built up with plywood and canvas superstructures to make them more seaworthy, although they often looked top-heavy on the water. An Eskimo with a job like Richard's lost time for subsistence hunting but often gained greater success in filling the ice cellar as the money bought good equipment—boats, snowmachines, fuel, weapons. At NAPA a brass shoulder gun sold for $2,000 and the casing for a single bomb to shoot from it was $50. When the *Plover* arrived in 1852, the Iñupiat hunted caribou with bows. When Lieutenant Ray's expedition arrived in 1881, rifles were providing the Iñupiat with more meat, but the caribou had become gun-shy and hunting with bows was no longer possible. The Iñupiat had become dependent on the outside world for ammunition, and there was no going back. In 2002, as the ice receded and the waves grew, safety and hunting success swung increasingly in favor of the bigger, faster boats. These were well out of the range of any full-time hunter whose cash income derived only from a few weeks of work here or there. Richard could have bought a car for the cost of the new outboard, and George Ahmaogak had two of them on his huge, extremely seaworthy aluminum boat, which he had loaned to Point Hope whalers for the season while he was running for reelection.

On this stormy day, two smaller boats joined with Savik Crew as we sped through the damp wind. One of them was Malik's odd-looking, partly home-built craft with its old motor, low gunwales, and high bow of plywood. Richard had chosen to launch from the boat ramp at Niksiuraq, at the end of the road partway out Point Barrow, where a sandy stub of land poked to the east into the shallow, protected embayment of five-mile-wide Elson Lagoon. His plan was to round the barrier islands and patrol the area to the north, but as we approached the far side of the lagoon the fog thickened and the three boats stopped and anchored. Savik Crew broke out the coffee and pastries. Roy asked Richard for a prayer. Roy started each day of whaling with a prayer, but he preferred not to do the out-loud praying himself. Richard was known as an eloquent prayer leader. The small group of men faced one another, heads bowed, backs to the wind-rough gray water. Richard prayed for strength, not success. He asked God for strength for each crew member of the kind each needed, as God alone knew. Then all were briefly quiet, in that small, pale space within the fog, looking at faces framed in parka hoods and thinking of what sort of strength the day might call for.

It was a long day and many things happened, but this next part I will always remember. Around midday, wind and fog turned back Savik Crew at a channel to the ocean and the boat anchored for shelter close inside the string of gravel bars known as the Tapkaluk Islands. Through the river of fog it was just possible to see white breakers crashing on the far side of the narrow barrier island. Presently, two objects appeared in the water to the west of us, a few hundred feet off. They looked like the heads of swimmers. Roy casually said they were a mother and a juvenile walrus and that they would approach the boat and try to get in because the boat was white and they would think it was an iceberg. I thought he was joking, but the swimming heads did near and circle around us. I could see the walruses' quill-like whiskers and huge, black tongues and hear their deep, resonant grunting as the young one climbed on the mother's back. Then we saw that a polar bear was pursuing the walruses, stalking them on foot along the beach of the barrier island. The back of its neck was smeared with dirt, making the bear look like a white horse with a brown mane. It was magnificent, peering at us with its neck raised and the parallel white lines of breakers at its back. The walruses were nearer to us, down in the water. The mother walrus, likely weighing a ton or more, raised her tusks in an aggressive show, then came to the stern of the boat to leap aboard. Roy poked at her with a long boat hook to fend her off. She circled to the side and Eben took the hook and poked her away again.

Malik came on the VHF and asked in Iñupiaq for a jump start. Without delay, Savik Crew raised anchor and powered up for the run west, to where Malik was anchored in rough seas off Plover Point. As we ran parallel to the seven-mile strand of the Tapkaluk Islands, I counted the polar bears standing along its beach, appearing from the fog and disappearing behind us. Thirty-five stood there on the barren gravel, evenly spaced as suited their unsociable nature, gazing at our speedy passage. Their spacing and their forlorn setting on that narrow strip, fog-bound, wind-whipped, and hundreds of miles from their home ice, reminded me of a shabby polar bear display—this was never how polar bear were meant to be seen. I preferred to think of them on the ice in the spring, stalking, swimming, and ruling. Then we reached Malik, came alongside, and looked down at his cheerful face on that little boat bouncing in the waves, the last time I saw him.

After September in Barrow, October weather in Princeton was glorious. Engineering professor Rob Socolow wore brown tweed and ecologist Steve Pacala shorts, a fleece pullover, and sandals as they strode across campus toward a meeting of their Carbon Mitigation Initiative team, a group set up to find a solution to climate change. Steve excitedly told Rob about data on wind farms. It seemed no one had ever studied the extent to which big groups of windmills slowed down the wind. If humankind were to supply a significant portion of energy needs from wind power, we would remove tremendous kinetic energy from the atmosphere—we would slow down the wind—and that could affect weather and perhaps climate in major ways. Steve wanted to do the calculations and discovered to his amazement that no one had ever done the basic research on which to base them. But today he had found a utility company manager who had measured the velocity of the wind on each side of a large wind farm, and that gave him what he needed to start.

"Why isn't this in a textbook?" he asked, and then answered his own question: "The people who work on wind energy like wind energy. The [alternative energy] area has been dominated by true believers on both sides, and what's needed is analysis by skeptics of both sides."

Rob founded the Carbon Mitigation Initiative group after being contacted by senior management at BP (the energy company), who wanted to fund an academic research team to find a path to a world beyond our current carbon dioxide emissions. Attacking the issue as an environmental problem and working with top people in the many related fields, Rob and Steve won $15 million from BP and another $5 million from Ford Motor Company. Two years of the ten-year program were over and the team needed to establish a strategy that could be the world's strategy, and then test it. Twenty brilliant minds convened in a lecture room to talk about the problem at the broadest scale, applying their knowledge of the potential solutions in the finest detail. The meeting, which unfortunately I could attend only on an off-the-record basis, sounded like a dialogue among the world's scientific philosopher-kings choosing the future of the human race and its effect on all life on earth.

Sunlight is the ultimate source of energy on earth except for that derived from the splitting or fusion of atomic nuclei or from the rotational momentum of the planet, such as the tides. The good news, Rob said, was that the sunlight striking the earth each day delivered ten thousand times more energy than humans used. The bad news was that we didn't yet have an efficient way to harvest it. Efficiency is the ratio of useful energy output for energy input: the efficiency of an internal combustion engine is 15 to 25 percent because that percentage of the energy in the fuel is converted to work, with the balance lost as heat and friction. Ecosystems are at most 1 or 2 percent efficient in capturing the sun's energy, and the effective efficiency from harvested plant matter is only about a tenth of a percent of the sunlight that originally hit the plant. Human beings were already using, probably, more than a third of all land-based organic photosynthesis just for food, fiber, firewood, and the like. Windmills and dams harvest the sun's energy more efficiently, but both take kinetic energy out of the natural environment and require significant changes to the landscape, and not many dams were left to build. Photosynthesis with machines such as photovoltaic cells take energy straight from sunlight, with efficiency in the laboratory as high as 30 percent, but significant technical challenges remained to producing efficient, inexpensive solar cells that could work for a long time without corroding. A power plant built with the photovoltaic cells currently available could produce electricity for a cost equivalent to $40-a-gallon gasoline. The environmental impacts would include mining for the exotic materials needed and covering large areas in permanent shadow—an area about the size of Colorado for the world's entire energy use.

In fifty years there may be a new technology that breaks through these barriers. Craig Venter, the maverick biology entrepreneur who sequenced the human genome faster and less expensively than the government, launched a project to build a living cell with a more efficient form of photosynthesis, a superplant that could cleanse carbon from the air and capture energy at the same time. To do that, his team would first need to invent both a chemical process for improved organic photosynthesis and a synthetic life form built of artificially encoded DNA to carry it out, two enormous breakthroughs. Scientists have predicted nuclear fusion could produce unlimited carbon-free energy from hydrogen, the most abundant element in the universe, in fifty years. I admit I was more hopeful about that prediction, however, when I read about it twenty-five years ago; I've grown to middle age, but fusion remains fifty years in the future. There are plenty of other big ideas, but they involve things like mining other planets or putting hundreds of huge photovoltaic power plants in space. All are fifty years or more away.

Unfortunately, we may not have fifty years. Steve Pacala said in fifty years it would probably be too late. Even if we started a crash program of new technology and aggressive energy conservation immediately, we would still be on track to a doubling of atmospheric CO_2 from levels prior to the industrial age. The scientists at Princeton, with their practical approach, were concentrating on bridging the time until new technologies became available. Only one energy source was cheap, abundant, and concentrated enough for immediate action: fossil fuels. Coal, oil, and natural gas are hydrocarbons; take out the carbon and you are left with hydrogen, which can be burned to produce energy without emitting greenhouse gases or, without any pollution, can be oxidized in a fuel cell that produces only electricity and water. Commercial processes already existed to separate hydrogen from carbon with efficiencies ranging from 60 percent for coal to 76 percent for oil. The infrastructure already existed to extract and distribute fossil fuels, and it could be modified to deliver hydrogen to cars and power plants instead. That new hydrogen infrastructure could be used again when renewable energy sources became cost-effective: solar, wind, or fusion could power extraction of hydrogen from water to run the fuel cells.

The problem Rob, Steve, and their colleagues had not settled was what to do with the leftover carbon after stripping it from the fossil fuels. The idea of dumping it in the deep ocean had been stymied by environmental objections; researchers couldn't even get permission to test the concepts involved. It would be chemically possible to make the carbon into artificial carbonate rocks, but the cost in energy would be high and the piles of rocks mountainous. The simplest and cheapest approach appeared to be to pump the carbon back into depleted oil fields. Oil producers already knew how to reinject waste, and carbon could even act as a solvent and field pressurizer to improve oil production. On a few fields where huge quantities of carbon dioxide came up as a petroleum byproduct, carbon was already being injected back into the ground. Deep injection might work in other geological formations and deep aquifers, too.

Part of the appeal of this concept was that oil companies could use it to make a lot of money and so might not resist the shift. To move the world to hydrogen fuel, governments would have to tax carbon emissions; the most popular version of such a system would allow carbon savers to sell tax credits to carbon emitters, creating a market in carbon credits that theoretically would encourage the most cost-effective carbon reduction projects. A test marketplace for trading greenhouse gas emissions, the Chicago Climate Exchange, began operation in 2003. Such a system would allow an oil company to sell a barrel of oil twice: once for the hydrogen energy contained in the hydrocarbons, and again for the carbon stripped out and sequestered underground. But Rob Socolow, the engineer, was preoccupied by the risk entailed in that transaction. What if the carbon leaked out? For the purposes of saving the atmosphere, some leaks would be acceptable, but companies wouldn't make the investment with the risk that their carbon credits could disappear into thin air.

Such problems seemed manageable, however, and the idea was promising enough to move the debate. As recently as the early 1990s, the energy industry and its political allies were united in denying climate change. Their representatives attacked the integrity of scientists speaking up about the issue in the political arena. By the mid-1990s, some companies, including BP, broke ranks and recognized the problem was real. As a Royal

Dutch/Shell executive said, "We don't want to fall in the same trap as the tobacco companies." Lord John Browne, BP's bright and appealing CEO, said that addressing carbon emissions saved the company money by reducing waste and helped put it in synch with the needs of society as a whole. Hard-liners such as Texas-based ExxonMobil and President George W. Bush continued in denial for another five years before having an apparent overnight conversion late in 2002. In November, skipping the step of admitting there was a climate change problem, ExxonMobil and a few other companies funded a $225 million carbon mitigation study at Stanford, dwarfing by an order of magnitude the Princeton program that researchers had previously considered very large. Within two months, Bush followed the same path when his 2003 State of the Union address called for a $1.2 billion program to develop vehicles propelled by hydrogen fuel cells. These moves made no sense absent climate change, which Bush and his Texas oil allies still denied as unproven.

On the basis of purely economic motives, however, the actions of ExxonMobil and the Bush administration did make sense. The president's scientific and environmental advisers had told him unequivocally that the greenhouse effect was real, and the oil executives in Texas surely knew it, too. The most profitable strategy, however, was to maintain the status quo as long as possible by demanding more and more scientific proof, while at the same time preparing a new business model for the day denial became untenable. Scientists eager for climate research funding played into the game nicely, especially modelers building predictive GCMs whose additional investments of time produced additional uncertainty. At the same time, by preparing for the inevitable switch to hydrogen fuels, ExxonMobil could avoid the fate of buggy whip manufacturers after the invention of automobiles. Eventually, reality would carry the day. The doubters themselves were already feeling the change. In Alaska, oil exploration was hindered by progressively shorter winter seasons, managers of the trans-Alaska oil pipeline were studying how to deal with melting permafrost, and the silos for Bush's missile defense initiative had to be designed with climate change in mind.

This question of sincerity preoccupied journalists. A framework of melodrama, separating good guys from bad, often surrounded reporting on the environment. ExxonMobil seemed to think it could escape being a bad guy by spending a lot of money, even without a sincere face like BP's

Lord John to say the right words. But writing big checks didn't help after the *Exxon Valdez* oil spill. The beach cleanup was feckless and often destructive, and the billions spent went over with the public like the church donations of a mafia don trying to buy his way into heaven. Money from dirty hands was suspect in climate change mitigation, too, and critics attacked the companies for making research grants while still exploring for oil, as if only funds given of pure heart could find good solutions. In the melodrama, climate change was a biblical test for our entire species: would we cast off our profligate ways in time to avoid fire, flood, and plagues of insects? What a letdown if ExxonMobil and BP solved it.

Rob Socolow, himself active in the Audubon Society, had run into the puritan branch of the environmental movement before, and now he was tangled up in convincing environmentalists that fossil fuels were not inherently evil. Some were steadfastly opposed to the idea of making hydrogen from fossil fuels and sequestering the carbon. Rob had met in Washington with leaders of several major environmental groups. One agreed to consider the concept in exchange for three concessions: the oil companies had to confess climate change was real, fossil fuel hydrogen couldn't displace renewable energy, and the oceans would be off-limits as artificial carbon sinks. Others wouldn't consider the idea on any grounds.

Why? A Greenpeace staffer in Alaska said using hydrogen from fossil fuels was like switching to low-tar cigarettes. The organization's executive director, John Passacantando, feared going a little way down that path could mean getting shanghaied into another century of fossil fuel use that would displace renewable energy. Instead, he favored turning to renewable energy and conservation immediately with natural gas (which emits relatively little carbon) as a stopgap in situations where renewable energy was not yet feasible.

"We're going to look back, whether it is twenty years from now or fifty years from now, and say, 'My God, that fossil fuel economy was madness, how could we ever have let eggheads at MIT say we couldn't get away from it,' " John said. "That's not a big prediction. In 1959, it looked like we would never have civil rights in this country."

Environmentalists were wise to call upon morality and rights in making the case for change, even at the cost of being dogmatic. Appeals to self-interest were more difficult to make. From the point of view of an individual, emitting the carbon dioxide necessary to drive to work or to fly to a Hawaiian vacation yielded a net benefit greater than the climate effects on that one person caused by that one action. Even summing up individual interests into national and global interests might not yield an unequivocal imperative to drastic action, because most of the people now living on earth would receive only partial benefits from reducing carbon emissions. The majority of the benefits would go to today's children and succeeding generations.

The difficulty is that once fossil fuel carbon is extracted from the earth's geology and inserted into the atmosphere and biosphere, it takes a long time to get back into the geology. Sedimentation at the bottom of the ocean is the only unlimited carbon swallower, and current emissions far exceed its appetite. The goal is to stabilize atmospheric CO_2, not reduce it, and to stabilize at any level would require fossil fuel emissions eventually near zero. Our actions to cut emissions can determine how much more CO_2 the atmosphere receives, but not what climate changes we live through. Even if we act drastically, the climate will keep warming during our lifetimes. The climate system hasn't caught up with the carbon dioxide we have already emitted. If we could stop emitting carbon dioxide immediately—an impossibility—the earth would likely still experience additional climate warming equal to what we have experienced so far. Our actions today have a much greater impact on people living after our deaths. Stabilizing the climate at a higher temperature in 150 years is an ambitious goal beyond our reach without major action as soon as possible. There are many unknowns, largest among them exactly how much the climate system responds to additional CO_2, but we can draw an outline beyond the uncertainties. Under highly optimistic but still credible assumptions, we would need to build a large power plant with zero carbon dioxide emissions every day from 2000 to 2050 to stabilize the climate at a level 2 degrees C higher by the year 2150.

Many scientists I met were incredulous that such a crash program hadn't started already. One simple reason it hadn't was that most people did not understand the problem as well as the scientists did. The diffusion of scientific and technical information was inefficient and haphazard. Senator Ted Stevens, a holder of the nation's purse strings and a concerned advocate for an area heavily affected by climate change, was unaware until I told him in October 2002 that hydrogen and carbon could be split and the carbon stored in oil fields like those on the North Slope. (Rob Socolow called Prudhoe Bay a nearly perfect site to sequester carbon.) How differently might Ted approach carbon trading knowing that Alaska could be one of its greatest beneficiaries?

The question remained, however, how the world would react once understanding was widespread. Humankind has never faced a similar choice. The lessons of evolution and ecology—all prior experience on earth—tell us nothing about an entire species giving up its cheapest and most abundant supply of energy for the benefit of unknown individuals living generations in the future. This goes beyond any social contract: a serious response to climate change would be fundamentally ethical. Those living 150 years in the future would have no way to reciprocate except by remembering our names, or cursing them. If that seems like incentive enough, consider all the religious, cultural, and national barriers between people. And then think of how hard it is in many American towns to get affluent local voters to support their own schools.

Steve Pacala, the Princeton ecologist, had no patience at all for my questions about whether humankind was capable of the ethical reach needed to save the climate. He accepted climate change as a personal challenge, talking and working as energetically as if the planet's fate depended primarily on him. Steve was the project's arm-waving idea man, a strategist quick to follow a train of thought wherever it led, his voice rising and falling over the emotional terrain of his intellect. At the meeting I attended, Steve got the conversation bubbling while his coleader, Rob Socolow, kept it from spilling over. Rob behaved as if he were on an important research project; Steve was grabbing at a chance to save the world. With the policy makers the project gave him access to, and his willingness to jump straight to the last square, that ambition didn't seem outrageous. He said, "I now believe that we could actually solve this problem in our generation. It wouldn't be finished, but we would be on a track that made it inevitable. And so I really have to try. It's a matter of responsibility."

Mainly, Steve thought of his teenaged children as he flew back and forth across the world—twenty-four hours for Europe here, three days for Japan there—and endured the antagonism of some of his fellow environmentalists. But a broader sense of obligation also drove him.

He said, "I believe that nature has enormous aesthetic value. I think that if we destroyed all the coral reefs in the world by letting CO_2 go too high, that that would be worse than burning down the Louvre on purpose, destroying all that great art. Whether or not it would have any functional significance for the world is a separate issue, but just as an aesthetic crime it ranks right up there. And I have no desire to be a vandal. So I think there is aesthetic and emotional content to the natural world that is undeniable.

"For instance, for me, I would like to see a lot of the Arctic left as a temple to wilderness. I went to ANWR last year, in part just to look at it with my thirteen-year-old son. And for me the aesthetic arguments over ANWR are much more compelling than scientific arguments over it. . . . If you get in a copter and fly from Prudhoe into ANWR, when you cross the river, you can tell. Any moron can tell you have gone from a place that is human-impacted to one that wasn't. And there's something deep within the soul that stirs when you look at a landscape that could have come straight out of the late Pleistocene."

In 1963, a storm hit Barrow in October, when the ice was far offshore. A surge of water destroyed fifteen houses and cut the road to NARL, where several buildings and the airport were wrecked and people had to be rescued by heavy equipment driving in deep water. The storm cut power, spilled much of the winter's fuel supply, contaminated the drinking water, and knocked out communications, including those used for transpolar flights. A lot of the shoreline washed away. A storm as bad as that has not occurred since, but big blows have come more frequently in the fall, powered by the change in the weather and able to build waves over hundreds of miles of open water, since the ice pack stays out farther now. After the 1986 storm that washed away the house sites of the frozen family and the buried shaman at the thousand-year-old ruins of Utqiagvik, the borough commissioned a study that, after another destructive storm in 1989, led to the construction of a custom-built $8 million floating dredge, the *Qayuuttaq*. By the mid-1990s, it was building up beaches to protect Barrow and Wainwright, but another unexpected fall storm in 2000 caught the dredge operators unprepared. Before they could move to the sheltered waters of Elson Lagoon, the *Qayuuttaq* was driven ashore and sank.

In the old times villages would simply move farther from shore. Captain Maguire of the *Plover* learned in 1854 that Nuvuk used to be a mile farther out, on a bar that was by then underwater. Today the last trace of Nuvuk, as remembered by Warren Matumeak and other elders, was disappearing into the waves. Missionaries pinned villages permanently to spots on the map around schools and churches, converting the Iñupiat's idea of a town to their own. In the process, the Iñupiat lost some of the adaptability to climate change that nature had imprinted on their culture.

In the early 1980s, the North Slope Borough used some of its new oil wealth to provide basic utility services all over town. The project, called the Utilidor, entailed building miles of insulated tunnels in the frozen ground at a cost of $270 million. With the current rate of erosion, a big storm like the one in 1963 might breach a low point in the Utilidor and flood it. The U.S. Army Corps of Engineers studied the idea of nourishing miles of beach with sand and gravel, raising the level of a coastal road to form a sort of seawall, and armoring it with concrete. Presumably, Ted Stevens would come up with the money.

Elsewhere in Alaska, the federal government was already paying for road repairs where a decade of warm weather had thawed the marginal permafrost. Large sections of a few long, two-lane highways that traversed the southern part of Interior Alaska, from the border with Canada to the coastal mountains, were sinking into mushy ground or getting buried under thawing mountains that slumped. The roads needed constant and repetitive reconstruction, forty miles in a summer. But the guys doing the work didn't regard it as a crisis. One said, "It's manageable. It's just a matter of money."

So it went in the richest state in the richest nation the world had ever seen. But even here, even with Uncle Ted's money, there were limits. Shishmaref, built on a barrier island in northwest Alaska, was losing property and improvements every year and had asked Ted Stevens for money to relocate. He was sympathetic but estimated that moving the ten Arctic coastal villages similarly threatened by climate change would cost around $100 million apiece, and that was more money than even he thought possible to obtain for populations of a few hundred people per village. He wanted less expensive choices. Shishmaref villagers feared that, without a move, their ancient community would disperse, as residents one by one abandoned a place where the threat of storms was becoming part of daily life.

Soon, the rest of the world would be coping with these issues. Barrow made a good test. Scientists began studying the community to see how it would handle the problem of climate change. Amanda Lynch, of the University of Colorado, came to town with a five-year, $2.5-million grant from the NSF to integrate the science of climate change with Barrow's adaptation to it. She allied with Anne Jensen locally to draw on Anne's membership in the community and her reputation, with her husband, Glenn Sheehan, for ethical research. Anne arranged meetings between the community and the team of experts in atmospheric science, coastal erosion, public policy, and other relevant fields to find out what local people wanted them to study. The project modeled changing storm tracks and ice position and the way the shoreline was responding, studied the earlier storms, and listened to elders' ideas about what made sense and what didn't. As an archeologist, Anne had worked on a seawall project in Point Hope years earlier, only to learn after the job was complete that the wall had been built in the wrong place. The middle-aged villagers who guided the project said storms came from the north, but the elders who remembered longer said the very rare storms that caused the greatest damage came from another direction. The scientists on the Barrow project hoped to find ways to avoid such mistakes. Their final product would be a book about their techniques that could be used in any community threatened by climate change.

Anne and her colleagues believed a strategic retreat from the shore would cost less and last longer than the Corps of Engineers' plan for a frontal defense against the ocean. Houses near the water could be moved to the far side of town as they came under threat, rolling Barrow slowly inland. Watertight doors could be installed in the Utilidor to contain flooding at any one location. New capital investment could be concentrated on higher ground. The people of Nuvuk used to absorb the blows of nature rather than fight them. As the sea devoured their houses and qargis, they dug new ones out of frozen ground with hand tools. "They probably invested a hell of a lot more work in those houses than people invest in their houses now," Anne said.

Of course, ancient people weren't tempted by the option of federal dredging and beach armor, which offered the prospect of staying in one place without adjusting to change. The Corps' plan seemed like a reasonable option from the vantage point of Oliver Leavitt's house, at the edge of a high ocean bluff next to ancient Utqiagvik. But Anne pointed out that as climate change progressed, Barrow might face a lot more competition to maintain and improve coastal defenses. Other towns would have more investment at stake and fewer options for relocation. The Corps wouldn't be able to dredge and build walls everywhere, and Ted Stevens wouldn't live forever. Prudence and Iñupiaq humility before nature seemed to counsel the path of retreat. "You think you're better than the weather? Let's see what the weather has got in store for you." Richard Glenn had said those words in another context, but they would make good general advice for the Corps of Engineers.

Climate change will probably alter society everywhere, but how? Many published predictions are as insubstantial as daydreams: one writer predicts social breakdown and world totalitarianism, another predicts a healthy boost in crop yields. You can predict anything you like by making choices on the many branching paths of uncertainty. But it is also possible, by discarding all but the most certain changes, to honestly estimate the basic challenges people will face. Sea level already has been rising since the last glacial period, when the Bering Sea floor was exposed and connected Alaska with Siberia. Sea level has been much higher in the past, too, as shown by ancient seashores far from the coast in the Arctic and many other places. Climate change accelerates the rise in sea level two ways: water expands when it warms, and melting continental ice increases runoff into the ocean (sea ice does not increase sea level when it melts, because it is already floating in the ocean). In the twentieth century, sea level rose four to eight inches, a rate ten times faster than the average for the previous three thousand years. Rising sea level, with increased erosion, storm damage, and flooding, will likely impose costs on many coastal communities. Each area's wealth will determine how it can respond. There will always be enough money to protect Manhattan Island, but most places will have hard choices to make, just like Barrow.

Ron Brunner, the University of Colorado public policy expert working on the Barrow study, saw these local decisions as the best avenue to a global climate change solution. When a town found ideas that worked, they would pass them on to other towns. Adaptation almost had to happen locally, and Ron thought carbon reduction could start there, too. Governor Jeanne Shaheen of New Hampshire signed carbon reduction legislation in 2002, noting that climate change "threatens skiing, foliage, maple sugaring, and trout fishing—all crucial to our state's economy." Many other U.S. states and local governments adopted carbon reductions based on local concerns, some more stringent than most national governments'. Companies such as BP and Johnson & Johnson were voluntarily cutting their carbon emissions for business reasons and because people wanted to do the right thing. If that seemed idealistic, Ron pointed out it might be the only real opportunity for progress. Social change could come either from coercion or from persuasion, and the top-down, coercive approach—scientists issuing reports and asking international organizations of governments to approve mandatory controls—wasn't working after decades of effort and $30 billion in research.

The problem wasn't that these scientists didn't understand the climate but that they didn't seem to understand people. Some climate modelers regarded their work as writing an "operator's manual . . . for Spaceship Earth." Ron wrote, "This implies a command center controlled by those few with the necessary technical expertise; the rest of us, presumably, are passengers along for the ride." But even if the scientists could produce the manual, no one was holding the steering wheel in the control center. That's not how the world works. In fact, there will be no operator's manual either, because the whole enterprise rests on a logical fallacy: events in an infinitely complex world, full of constantly adapting people and natural systems, cannot be predicted reliably by a mathematical code. Understanding climate change, as well as responding, must happen inside individual human beings, in their minds and in their bones, through judgment and trial and error, in the way the Iñupiat, and all people, learn the truth by living it. We needed modern science, Ron wrote, but we also needed a ten-thousand-year-old science based on the human experience of concrete places and events.

The solution was simpler than many people wanted to accept. Ron said, "Not enough people in the world, and particularly their leaders, want to do anything about this. So what do you do about it? You start with the easy cases, the people who already care and want to do something." Communities that felt compelled to act would influence one another. That, he hoped, would eventually lead to broad social change, a building of pressure upward from the towns and cities that would force governments to react nationally and globally to reduce carbon emissions. In the usual way social change occurs, ordinary people would communicate, their ideas and reactions spreading from one town to the next, to cities, to nations, and across cultures until the world was ready for action.

I asked if we had time for broad social change. Ron said, "No one can know that. But what's the alternative?"

Climate change that happens gradually is difficult for people to perceive. Even in Barrow, where the Iñupiat depended on wildlife, ice, and the timing of the seasons for their livelihood, and where scientific lectures were a popular form of evening entertainment, men like Oliver Leavitt and Richard Glenn became believers only during that terrible spring whaling season of 2002. Would a few cold winters revive their doubts? Most scientists weren't much help. Uncertainty was their daily bread. Not enough had Suki Manabe's ability to rank important certainties above trivial unknowns. The news media picked up this noise and amplified it, covering minor scientific controversies without context or scale and whipsawing readers back and forth in the same way they did with dietary advice: salt was bad for you, then it was good; the world was warming, then it wasn't.

The global profile of the problem was distant and depressing, like so many other environmental crises that came and went. (On New Year's Day of 2000, several of my friends shared a sense of relief that the world had survived; I wonder how many others who grew up amid the pessimism of the 1970s felt the same way that day.) Besides, said Amanda Lynch, any statement about the entire globe was fundamentally abstract—hard to quantify and, once quantified, hard to understand. On the other hand, she said, clear and specific statements could be made about particular places. She already had the tools to make firm predictions of the probability of different kinds of storms and the damage they would cause in Barrow over the next fifty years. That was something that people could get their minds around.

Anne Jensen said, "People here see that things are changing, but I think that urban people don't—which is most of the people in the world—because they're so encapsulated. I think a lot of people are disconnected from climate because they're so insulated from it."

She had been away from Philadelphia and Bryn Mawr long enough that she sounded like an Eskimo in her disdain for people who ate veal but condemned hunting. Yet those were the people who would have to feel climate change coming and develop a personal and political desire to address it. Three-quarters of Americans lived in cities; any major social movement would have to be mainly urban. When that movement comes, perhaps the

Iñupiaq worldview can help others to see. That is the dream implicit in Anne, Ron, and Amanda's work: a cultural transformation that allows many people, as one, to see the natural world as it really is.

The fall whaling season was plentiful. A large cargo plane carried whale meat from Barrow to Point Hope. Tables were well supplied at Thanksgiving and Christmas. Savik Crew did not get a whale. Richard Glenn missed much of the season attending the Alaska Federation of Natives conference in Anchorage. But he had a good winter, as the dark part came with preparations for Kivgiq, the Messenger Feast.

Kivgiq died out for a long time. The last interior Iñupiat village, people from the Colville River, perished sometime after 1910 as its members tried to get home from Kivgiq in Barrow. They caught the measles from sailors and, despite their being desperately ill, their shamans counseled them to make the trek home, upriver. Charles Brower found their bodies and Kivgiq gifts on the riverbanks. The last traditional Kivgiq was held in 1915, but the dancing and gift giving survived later in the century, moving to Christmas, a similarly joyous gift-giving holiday. Then, in 1988, Mayor George Ahmaogak called together a formal Kivgiq once again and people came from villages all over Alaska's Arctic. The festival, modified from earlier days but with the same spirit, was too expensive to hold every year, so the celebration in February 2003, with invitations to seven hundred dancers, came with great anticipation.

In the 1980s, Warren Matumeak and Walter Akpik, Arlene's grandfather, had started a dance group for the mayor called Suurimmaangitchuat, meaning, "Those who don't really care," although Richard joked that it could be called "Those with the longest sign" or "Those who used a random letter generator to come up with their name." Walter was aging and the group had not performed in six years. Warren, a talented composer, asked Richard and Arlene to help revive the group so its songs and dances would survive into the future. Through the early winter they practiced every day when not prevented by church or basketball: drummers, male dancers with their powerful, lunging strides, and female dancers, whose feet stayed planted as they moved gracefully with their hands and by slightly bending their knees. Elders, young people, and children danced side by side, learning through the traditional process of watching and then joining in. Richard, whose life so often pushed him into positions of authority, found special joy in joining a group without stars—everyone was equally responsible to drum in time and sing in tune.

Some dances told a story and others invited everyone to join the dance together. For those, the audience in the gymnasium would come down from the bleachers and dance along with the group. Richard and Arlene practiced one of Warren's dances to perform alone as a couple. It made a good winter, having this work to do together. The words were simple, a chant that meant, "Let's be happy and go to their land," repeated in high spirits with stylized dancing, a come-join-the-fun dance. The audience loved it. The festival went on for three days.

In the federal budget that passed that winter, Ted Stevens sent $1 million to plan the new Barrow Climate Change Research Center. As part of a program for minority contractors, the money was steered to UIC Construction, a subsidiary of the village corporation that employed Anne Jensen. Together, BASC and UIC would try to get "as much as possible in the ground," Richard said, so that the next phase of funding couldn't be swiped for another purpose.

Matthew Sturm, Jim Maslanik, and the aerosonde guys from Australia returned in March for their project testing the vision of NASA's new microwave satellite. It was a cold month in Barrow, far below zero degrees F most of the time, but April Cheuvront envied them anyway from her vantage on her mountain in the springtime of North Carolina. She had spent some time in the snow that winter, a week taking measurements with Glen Liston in Colorado, but on that trip she slept in a bed and showered every night, so it wasn't like the cross-Alaska transect. A substitute covered for her while she was gone. "I have a really good principal, and he realizes I need to do these things," April said.

April still lingered in front of her future like a girl at the top of a slide. She and Steve were looking at log house plans and getting ready to put in a road to their property that summer. She was looking forward to having babies and being a stay-at-home mom. But she also had planned a cross-country drive to Alaska with one of her teacher friends during the month of June, going all the way in her Toyota pickup, visiting the gang in Fairbanks, and backpacking in Denali National Park and the Brooks Range. Steve couldn't take that much time off from his law practice, so he would fly up for a week of fishing. April sounded torn, like an Alaskan trapped in a North Carolinian's body, as she speculated about what other choices she might be able to pursue when her children, not yet born, were grown up and gone. She would like to be a scientist and to live in Alaska, but she didn't want to leave her extended family or defer the family she planned to start. She said, "I'm just going to kind of wait and see where life takes me. If it takes me to Alaska, then we'll do it, but it just didn't seem right this time." Four days into her trip, April learned she was pregnant.

Once, you were either an Eskimo hunting whales or a southern woman in a schoolroom. Now you could choose a place and choose how to live in it. Cultures with their different ways of being in nature bumped up against each other, mixed, and were newly created. But nature was one, and therein lay the responsibility that came with this new freedom. When culture became a choice, an individual could no longer claim to be merely its creature, driven by the dictates of an ideology given at birth. You were obliged to examine, at a deeper level, who you were, what you were, and where you fit in with the living things around you. Fossil fuels helped give us this physical and social mobility. If that freedom allows us to see nature and our predicament more clearly, we might find a way to avert some of the damage fossil fuels have caused. The problem is cultural, and so is the solution.

As April found, however, once you checked with yourself, it could be that selecting a culture and a home was much harder than you thought. These were heavy doors to open. But there was inspiration in the cultural journeys made by Richard Glenn and Kenny Toovak. Richard, a child of Silicon Valley, just short of a doctorate in sea ice, became a whaling captain and preacher of Iñupiaq humility. Kenny, whose main schooling came from his father's lessons about hunting and humility, became a beloved icon of Arctic science. In May 2003, the symmetry of their lives was completed. Richard and former NARL director John Kelley prevailed in their letter-writing campaign to the University of Alaska Fairbanks for Kenny to be awarded an honorary doctorate. And the dance group Richard and Arlene had joined would lead the procession at the commencement ceremony.

The 2003 commencement came during the heart of spring whaling. Savik Crew built a trail and a camp far south of town, in the area Roy Ahmaogak had always wanted to try, but after another warm winter the ice was bad and changeable and they kept retreating to avoid break-off conditions and a big ivu. With the time approaching for Richard to leave for the commencement in Fairbanks, two other crews caught whales at the spot Savik Crew had abandoned. "We started that old sport, we started that doubting game," Richard said. But presently the lead where they camped filled with more spouts than they could count, perfect conditions. Roy steered the umiaq and Eben threw the harpoon; although he hit bone and the darting gun did not go off, the harpoon tip held and the whale towed a buoy. Richard pursued in the aluminum boat and killed the whale, a choice forty-three-foot, ten-inch male, the crew's first in three years. Savik himself carried the flag back to town, on a snowmachine, holding it high, reenacting his favorite story, a tale from his younger years when he ran the flag home and beat on the outside of the house with a two-by-four, a story he had told to Richard and Roy so many times they could move their lips with the words. Now the yard again contained a huge pile of whale meat and the flag again flew high.

Richard and Arlene and the Suurimmaangitchuat dancers left for Fairbanks before Savik Crew had fed the community, but still they were late for a reception for Kenny on a sunny patio at UAF's Iñupiat dorm and commons. Three former NARL directors were there and the chancellor of the university, all paying Kenny tribute. Glenn Sheehan also spoke, saying, "The traditional knowledge of Barrow includes science." Kenny's speech, with his deep, strong voice and his storytelling gift, upstaged them all, as he graciously turned credit for his accomplishments to other Iñupiaq science workers, to the scientists, and to his late wife, Thelma. Tears welled up in his eyes. "All the rest of you people. I didn't go any further without any help. I can't go any further without help. From the Natives of the North Slope. Without help, alone, a man alone, I can't go any further."

Richard, Arlene, their daughters, Warren Matumeak, and the rest of the dancers arrived in rented cars wearing their matching bright purple *atikluks* (cloth summer parkas), and the festivities began, with drumming, dancing, and food, including little hunks of maktak. Richard showed around his eleven-month-old daughter, Joanne, in a tiny atikluk, telling Max Brewer, the old NARL director who sent Arlene's father to the ice island, "The princess has another helper." At first only members of the troupe danced on the concrete patio, but Warren teased and coaxed the audience and soon pulled the other Iñupiat into the dancing, and then the retired scientists began grinning and giving in to invitations to dance, and finally Kenny, stiff with age and dignity, joined in with a great, happy smile.

Next day, in the lobby of the sports center where the commencement ceremony would be held, the Suurimmaangitchuat dancers' purple atikluks mixed nicely with the bright medieval robes worn by hundreds of milling academics. Kenny, nervous and hot in a dark pinstriped suit, was escorted into a cloakroom reserved for the most distinguished dignitaries, where women draped him in a gown and put a mortarboard cap on his head, like any graduate. They were already addressing him as Dr. Toovak. The dancers led the procession through the huge gymnasium, beating their drums slowly and chanting their most ancient song, the traditional processional for Kivgiq, "as old as Eskimo dancing," Richard said. At first inaudible under the buzz of the arena's lights, the sound of the drums and voices grew as the troupe approached the dais, a tiny band of men, women, and children in an enormous sea of faces and chairs. Warren and Richard led the way up onto the platform, followed by Arlene, baby Joanne and the girls, and the other members of the group. They briefly faced the audience, then they marched up the stairs to sit in an upper balcony, far above the stage, where they could look down upon Kenny taking his seat of honor.

Notes

PREFACE

xii Average winter temperatures in Interior Alaska: National Assessment Synthesis Team, U.S. Global Change Research Program, *Climate Change Impacts on the United States: The Potential Consequences of Climate Variability and Change; Overview* (Cambridge: Cambridge University Press, 2000), p. 74.

1. THE WHALE

4 Billy Jens Leavitt: He pronounces Jens with a hard J, like jam, not as it is pronounced by the Danish.

6 1961 closed duck hunting and duck-in: Oliver Leavitt interview; Henry P. Huntington, *Wildlife Management and Subsistence Hunting in Alaska* (Seattle: University of Washington Press, 1992), pp. 28, 42–43; Joel Gay, "Feds Plan Limited Legalization of Subsistence Bird Hunting," *Anchorage Daily News*, February 14, 2003.

9 the big ivu in 1957: Analysis of this event is in Craig George, Karen Brewster, et al., "Iñupiat Hunters and Shorefast Ice: How a Dynamic System Can Surprise Those Who Use It," submission to *Arctic*, in draft.

9 "Right after it had bit Aanga": Transcribed in Iñupiaq with side-by-side translation in *Puiguitkaat: The 1978 Elders' Conference* (Barrow: North Slope Borough Commission on History and Culture, 1981), pp. 394–95.

10 Some handy words: *Abridged Iñupiaq and English Dictionary*, compiled by Edna Ahgeak MacLean, a joint publication of the Alaska Native Language Center, University of Alaska Fairbanks, and the Iñupiat Language Commission, North Slope Borough, Barrow, 1980, pp. 26, 66, 69.

10 For example, *pigña*: For this example and much of what I have to say about Iñupiaq,

I am indebted to Dr. Edna Ahgeak MacLean, president of Ilisagvik College and author of the definitive dictionaries of the language, whom I interviewed on April 30, 2002.

15 When Craig arrived: Ray Dronenberg, "*Alumiak* and *Natchik*: An Account of Shipwrights and Seafaring at NARL," in *Fifty More Years Below Zero: Tributes and Meditations for the Naval Arctic Research Laboratory's First Half Century at Barrow, Alaska*, ed. David W. Norton (Fairbanks: Arctic Institute of North America and University of Alaska Press, 2001), p. 237. This dictionary-sized volume is a loving collection of scientists' memories and discoveries from Barrow's NARL era.

16 a population estimated by government scientists: The range of the estimate was 800 to 2,000. Thomas F. Albert, "The Influence of Harry Brower, Sr., an Iñupiaq Eskimo Hunter, on the Bowhead Whale Research Program Conducted at the UIC-NARL Facility by the North Slope Borough," in *Fifty More Years*, p. 266.

17 Commercial whalers: I was assisted with whaling history by John Bockstoce, whom I interviewed on December 3, 2002.

17 A paper delivered to an IWC conference: Howard W. Braham and Jeffrey M. Breiwick, "Projections of a Decline in the Western Arctic Population of Bowhead Whales," paper presented to the IWC Scientific Committee, June 1980; IWC reference number SC/32/PS8.

17 Some environmentalists, including: Henry P. Huntington, *The Alaska Eskimo Whaling Commission: Effective Local Management of a Subsistence Resource*, master's thesis, Scott Polar Research Institute, University of Cambridge, 1989, pp. 15–16, 19.

17 The North Slope Borough helped fund: Interview with Maggie Ahmaogak, September 23, 2002.

18 Other whalers trusted Tom: Albert, "Influence," 265–70.

21 one whale was 211 years old: John Craighead George et al., "Age and Growth Estimates of Bowhead Whales (Balaena mysticetus) via Aspartic Acid Racemization," *Canadian Journal of Zoology* 77, no. 4 (1999): 571–80.

21 the bowhead population was strong and rising: Albert, "Influence," p. 273. The IWC accepted the 7,200 figure based on the 1985 data in 1987. The same data was later reanalyzed to produce a figure of 7,800. Statistical confidence in each of these figures is plus or minus more than 2,000.

22 (and $10 million): Albert, "Influence," p. 270.

22 Better studies conducted in 1996: Interview with Tom Albert, September 10, 2001.

24 "down coast by Will Rogers monument": The Will Rogers Monument marks the place where Rogers and his pilot, Wiley Post, perished in a plane crash in 1935.

27 A foot-deep snow pack had disappeared: Based on the record kept for the Barrow Wiley Post–Will Rogers Airport, available online from the NOAA National Climate Data Center at www.ncdc.noaa.gov.

28 the snowmelt date had gotten eight days earlier: Climate Monitoring and Diagnostics Laboratory Summary Report No. 25, 1998–99, National Oceanic and Atmospheric Administration, Boulder, 2001.

29 A whale this size weighs: Craig George has developed bowhead weight estimates

based on length. A forty-foot whale weighs about twenty tons, a fifty-foot whale fifty tons, and a sixty-footer eighty to a hundred tons. In the United States, tractor trailers are limited to forty tons.

29 One was killed instantly: Based on interviews with Craig George and Richard Glenn, and on the *Anchorage Daily News*, "2 Women Die as Barrow Lands Whale," May 31, 1992, by the Associated Press.

30 She was probably fifty to one hundred: Craig George offered this rough estimate. Samples from each whale's eyes are sent for dating by the aspartic acid method, but results take years to get back to Barrow.

2. THE IÑUPIAT

34 the First International Polar Year: For the International Polar Year in general, see William Barr, *The Expeditions of the First International Polar Year, 1882–83: The Arctic Institute of North America Technical Paper No. 29* (Calgary: University of Calgary, 1985). I also used Clive Holland's *Arctic Exploration and Development c. 500 B.C. to 1915: An Encyclopedia* (New York and London: Garland, 1994).

36 "The Natives": Lt. P. Henry Ray, *Report of the International Polar Expedition to Point Barrow, Alaska* (Washington: Government Printing Office, 1885), pp. 22–23. I have also drawn on John Murdoch, *Ethnological Results of the Barrow Expedition* (1892) (Washington: Smithsonian Institution Press, 1988), which includes Murdoch's personal letters and an informative foreword by William W. Fitzhugh.

37 "Many of the old conservative men": Ray, *Report*, p. 48.

38 Greely obeyed prearranged orders: For a readable account of Greely's ordeal, see Alden Todd, *Abandoned: The Story of the Greely Arctic Expedition, 1881–1884* (1961; repr., Fairbanks: University of Alaska Press, 2001). Todd also assisted me in personal communication. A couple of points in the paragraph are matters of historical debate. Barr (*Expeditions*, p. 34) argues that Greely could have saved his party if he had learned from the Inuit of year-round ice-free waters in Smith Sound and how to get around them. Fitzhugh (Murdoch, *Ethnological Results*, p. xxviii) speculates that Greely's difficulties led to Ray's early recall.

38 According to local legend: I have been unable to verify this version of the purpose of the pit from original documents. The story is told this way in *The Future of an Arctic Resource: Recommendations for the Barrow Area Research Support Workshop* (Fairbanks: Arctic Research Consortium of the United States, 1999).

39 The scientific legacy of the International Polar Year: Barr, *Expeditions*, p. 206.

39 2007 International Polar Year: Initial planning meetings were held in 2002 in several countries. The National Academies were helping coordinate U.S. participation (dels.has.edu/ipy).

39 use of 1880s meteorological data: Interview with Roger Colony, UAF-IARC, March 25, 2002.

39 measuring Barrow more extensively: Bernard Zak, of the Sandia National Laboratories, made this statement at a talk in Barrow on April 27, 2002. Zak has visited the world's other Arctic research sites; he said Svarlbard, Norway, is a distant second.

40 put living Alaska Native human beings on display: An exhibit in 1996 at the Aluutiq Museum in Kodiak, Alaska, documented the display of Aleut people at the St. Louis World's Fair of 1904.

41 Before researchers could talk to elders: This requirement of King Island villagers was reported at a seminar October 19, 2001, at the annual conference of the American Folklore Society in Anchorage.

41 The Medieval Warm Period: For the impact of the Medieval Warm Period and the Little Ice Age on Old World societies, I drew on a readable book by Brian Fagan, *The Little Ice Age: How Climate Made History, 1300–1850* (New York: Basic Books, 2000). For the paleoclimatology of the period, Wallace S. Broecker, "Was the Medieval Warm Period Global?" *Science*, vol. 291, no. 5508 (February 23, 2001): 1497; and Thomas J. Crowley, "Causes of Climate Change over the Last 1000 Years," *Science*, vol. 289, no. 5477 (July 14, 2000): 270; and Feng Sheng Hu et al., "Pronounced Climatic Variations in Alaska During the Last Two Millennia," *Proceedings of the National Academy of Sciences* 98, no. 19 (September 11, 2001): 10552–56.

42 Birnirk people hunted: Don E. Dumond, *The Eskimos and Aleuts*, rev. ed. (London: Thames and Hudson, 1987), pp. 128–50.

42 Around 1000, the Thule whaling tradition: Ibid., p. 145.

42 the Thule displaced or assimilated earlier Canadian peoples: Ibid., pp. 93–98.

42 Poor-quality Alaskan pottery: Glenn Sheehan explained the Thule expansion to me and gave me some of these details. The theory was advanced originally by Robert McGhee, and he explains his ideas in several popular books, including *Ancient People of the Arctic* (Vancouver: UBC Press, 1996) and *Ancient Canada* (Hull, Que.: Canadian Museum of Civilization, 1989).

42 "Hey, you talk like our grandparents": Richard Glenn told me this incident happened to him.

43 Inuit paddling kayaks showed up: Fagan, *Little Ice Age*, pp. 113–16.

43 The Thule way of life could not continue: Dumond, *Eskimos*, p. 147.

43 people had to congregate: These theories and the evidence behind them are contained in Glenn Sheehan, *In the Belly of the Whale: Trade and War in Eskimo Society* (Anchorage: Alaska Anthropological Association, 1997), which is a revised edition of his doctoral dissertation.

44 Not a system of voting: Captain Rochfort Maguire of the British Admiralty and other early visitors observed this system of authority by consensus. Counterexamples exist, such as the brutal umialik Attungowrah described by Charles Brower, but Glenn Sheehan suggests he may have been an anomaly of the stressed, postcontact era, essentially put into power by visiting whites. In spirit, the consensus system of authority holds true in Barrow today, except that whaling crews now are much smaller, generally built around a family core. Maguire found just two active crew groupings in Nuvuk when he arrived.

44 seaside villages grew: Sheehan interviews and *In the Belly*, pp. 1–3.

45 Each whaling group building its own *qargi*: The last qargi was demolished early in

the twentieth century; Glenn Sheehan excavated one at Utquagvik in 1982, the only such dig to have been conducted (Sheehan, *In the Belly*, p. 9).

45 Its corollary today: This idea was given to me by Richard Glenn.

45 many words for "meet": *Abridged Iñupiaq and English Dictionary*, compiled by Edna Ahgeak MacLean, a joint publication of the Alaska Native Language Center, University of Alaska Fairbanks, and the Iñupiat Language Commission, North Slope Borough, Barrow, 1980, p. 127.

46 half the annual precipitation of Tucson: Western Regional Climate Center (www.wrcc.dri.edu).

47 Coming back from breakfast on Good Friday: This account is based on newspaper accounts at the time as well as interviews with Glenn Sheehan and a visit to the farmhouse in Conshohocken in October 2002. The articles include, from the *Philadelphia Inquirer*: "2 Survivors Recount Day of Carnage" (June 28, 1988), "Faulkner Witnesses Ruled Out" (December 3, 1988), "Negligence Trial Opens in Slayings" (July 25, 1989), "Montco Murder Suit Settled out of Court" (July 27, 1989), "Deal Would Reverse Man's Death Sentence" (July 3, 2002). And from the *Philadelphia Daily News*: "2 Slain, 2 Injured in Montco Stabbings" (April 2, 1988), "Suspect's Brutal 1984 Rape Try" (April 5, 1988), "Salvation Army Sued in Slaying" (May 6, 1988), "Voices Drove Him to Kill, Defendant Told Psychiatrist" (December 5, 1988).

47 Anne told the surveyors: This aspect of the story was not reported in the newspapers.

49 fifty prehistoric house mounds: Besides interviews with Sheehan and Anne Jensen, this section is based on several published accounts, including Bill Hess's spectacular book, *The Gift of the Whale: The Iñupiat Bowhead Hunt, a Sacred Tradition* (Seattle: Sasquatch Books, 1999), and the following newspaper articles from the *Anchorage Daily News*: "Treasures Destroyed" (September 28, 1986), "Erosion Reveals Prehistoric Body in Cliff" (August 5, 1994), "Scientists Use Fire Hose to Free Body in Ice" (August 11, 1994), "Eskimo Girl's Body Holds Glimpse of Past" (August 23, 1994), "Barrow Case Sets Archaeological Standard" (February 10, 1995), and "800 Years Later Remains Tell a Story" (March 25, 1995). And from the *Philadelphia Inquirer*: "An Icy Grave Yields a Look into the Past in Alaska" (October 23, 1994) and "Lessons from an 800-Year-Old Corpse" (March 14, 1995).

50 the Iñupiat paid aged and disabled elders utmost respect: Ray, *Report*, p. 44. Dr. John Simpson noted the same thing in his 1855 "Observations on the Western Esquimaux and the Country They Inhabit," published in *The Journal of Rochfort Maguire, 1852–1854*, vol. 2, ed. John Bockstoce (London: Hakluyt Society, 1988). Also writing at a time of starvation, he states: "For the tender solicitude with which their infancy and childhood have been tended, in the treatment of their aged and infirm parents they make a return which redounds to their credit, for they not only give them food and clothing, sharing with them every comfort they possess, but on their longest and most fatiguing journeys make provision for their easy conveyance. In this way we witnessed among the people of fourteen summer tents and as many boats, one crip-

pled old man, a blind and helpless old woman, two grown-up children with sprained ankles, and one other old invalid, besides children of various ages, carried by their respective families, who had done the same for the two first during many successive summers" (p. 520).

52 the attribution error: David G. Myers, *Intuition: Its Powers and Perils* (New Haven: Yale University Press, 2002), pp. 110–13; and Myers, *Social Psychology*, 7th ed. (New York: McGraw-Hill, 2001), pp. 90–91.

52 The people of Nuvuk defended their village: Introd., *Journal of Rochfort Maguire*, pp. 5–6.

53 the *Plover*'s stay at Nuvuk: All of this is based on Maguire's journal, Simpson's monograph, and Bockstoce's introduction to them; individual incidents can be found using Bockstoce's excellent index.

54 "He tasted almost every dish": *Journal of Rochfort Maguire*, p. 321. The passage is probably his quotation of Simpson's journal.

54 When Lieutenant Ray arrived at Point Barrow: This combines Maguire's journal, p. 130, n. 1, and Ray, *Report*, p. 44.

54 "In all their intercourse": Murdoch, *Ethnological Results*, pp. 52–55.

55 by the time they reached as far as Barrow, in 1854: Ibid., p. 53.

55 killed more than 7,000 whales: John R. Bockstoce, *Whales, Ice, and Men: The History of Whaling in the Arctic* (Seattle and London: University of Washington Press, 1986), p. 346. Bockstoce is the preeminent historian of Arctic whaling; I relied on his books and an interview with him.

55 When explorer Vilhjálmur Stefánsson surveyed the coast: Vilhjálmur Stefánsson, *My Life with the Eskimo* (1913; repr., New York: Collier Books, 1962), p. 71.

55 the death of the last Iñupiat interior village: Charles Brower, *Fifty Years Below Zero: A Lifetime of Adventure in the Far North* (1942; repr., Fairbanks: University of Alaska Press, 1994), pp. 228–29. Glenn Sheehan told me this was the last interior Iñupiat village. Details also come from a document held by the North Slope Borough School District: Chris B. Wooley and Rex A. Okakok, "Kivgiq: A Celebration of Who We Are," presented to the 16th Annual Meeting of the Alaska Anthropological Association, March 3–4, 1989, Anchorage.

55 The Eskimos' shore-based whaling also had advantages: Bockstoce, *Whales, Ice, and Men*, pp. 152–59.

56 Leavitt and Brower caught only one whale: Brower's description is in *Fifty Years Below Zero*, pp. 99–125. The story of Iñupiat and Yankee whaling also is well told in Bockstoce's *Whales, Ice, and Men*, pp. 231–54. Bockstoce tells me he checked the account in Brower's journals and manuscript carefully and found it highly accurate, although the published version of *Fifty Years* was heavily edited and sensationalized. Nonetheless, I recommend Brower's book: you can't put it down.

56 commercial whaling disappeared in 1907: Bockstoce, *Whales, Ice, and Men*, p. 335.

57 New England whaling heritage in Eskimo families: A few years ago, Oliver Leavitt and Mayor George Ahmaogak persuaded Senators Ted Stevens and Ted Kennedy to

cosponsor legislation creating an alliance through the National Park Service of the Iñupiat Heritage Center and whaling historic sites in New Bedford, Massachusetts.

59 Besides, science couldn't say: This paragraph is based on a lecture Philander delivered at Princeton on October 15, 2002. If Philander is right, we will know because El Niño weather patterns will become permanent.

60 Climate changes of the current interglacial period: Broecker, "Was the Medieval."

60 temperatures are probably sweeping upward faster: Jan Esper, Edward R. Cook, and Fritz H. Schweingruber, "Low-Frequency Signals in Long Tree-Ring Chronologies for Reconstructing Past Temperature Variability," *Science*, vol. 295, no. 5563 (March 22, 2002): 2250.

3. THE SNOW

62 Glen's computer model: Glen E. Liston and Matthew Sturm, "A Snow-Transport Model for Complex Terrain," *Journal of Glaciology* 44, no. 148 (1998).

63 Later a colleague told me: Carl Benson made this remark. Matthew himself rejects it, saying others deserve the title equally.

64 But automation certainly wasn't the answer: This paragraph is based on interviews with Sturm and Liston, and on Glen E. Liston and Matthew Sturm, "Winter Precipitation Patterns in Arctic Alaska Determined from a Blowing-Snow Model and Snow-Depth Observations," *Journal of Hydrometeorology* 3, no. 6 (December 2002): 646. The snow gauge issues are covered in Carl S. Benson, "Snow and Ice Research Begun at and Continued from NARL," in *Fifty More Years Below Zero: Tributes and Meditations for the Naval Arctic Research Laboratory's First Half Century at Barrow, Alaska*, ed. David W. Norton (Fairbanks: Arctic Institute of North America and University of Alaska Press, 2001), pp. 151–61.

71 A Barrow biologist and musher: Geoff Carroll told of his 1985 expedition from the north tip of Ellesmere Island at a lecture at BASC on September 21, 2002. A book was published about the expedition: Will Steger with Paul Schurke, *North to the Pole* (New York: Times Books, 1987).

73 the Barrow snow gauges were missing most of the snow: Benson, "Snow and Ice Research."

78 the raw data was likely bound for oblivion in an archive: The data was archived at the National Snow and Ice Data Center in Boulder, Colorado, but Sturm said it was unlikely to be used by scientists unrelated to the transect.

78 an answer to the question posed by Carl Benson: Liston and Sturm, "Snow-Transport Model."

81 Less than 10 percent: North Slope Borough 1998–99 Economic Profile and Census.

82 studying how snow and brush relate: Matthew Sturm et al., "Snow-Shrub Interactions in Arctic Tundra: A Hypothesis with Climatic Implications," *Journal of Climate* 14, no. 3 (February 1, 2001); Glen E. Liston et al., "Modeled Changes in Arctic Tundra Snow, Energy, and Moisture Fluxes Due to Increased Shrubs," *Global Change Biology* 8, no. 1 (January 2002).

82 They quickly published a piece in *Nature*: Matthew Sturm, Charles Racine, and Kenneth Tape, "Increasing Shrub Abundance in the Arctic," *Nature* 411, no. 6837 (May 31, 2001).

4. THE LAB

87 the AMSR-E: Claire Parkinson, NASA's project scientist on the Aqua satellite at the Goddard Space Flight Center, provided information on the satellite, as did the project Web site, aqua.nasa.gov. Maslanik's research proposal to Goddard, *Validation of AMSR-E Polar Ocean Products Using a Combination of Research and Modeling*, contains a discussion of the weaknesses of various algorithms. I have also drawn on information from a project called SHEBA (Surface Heat Budget of the Arctic Ocean), of which Maslanik's 1998 test was a part (see note to p. 93, below).

90 Jim, Matthew, and their team met with Warren: In the interests of the story, I have combined events from two similar outings on two days and made insignificant alterations to the chronology.

93 It would take some basic information: This discussion is based on several sources, including Hajo Eicken, of the University of Alaska Geophysical Institute, and Donald K. Perovich, of CRREL. Perovich helped oversee SHEBA (Surface Heat Budget of the Arctic Ocean), which involved freezing an icebreaker in the ice pack for an entire year of measurements, producing solid numbers for ice albedo and heat flux in various conditions. I heard him speak and interviewed him in Hanover, New Hampshire, in October 2002 and relied on a paper he coauthored: Teneil Uttal et al., "Surface Heat Budget of the Arctic Ocean," *Bulletin of the American Meteorological Society* 83, no. 2 (February 2002). Eicken provided an understanding of the outstanding problems of the field in an interview in Fairbanks in July 2002 and in a book chapter he coauthored with Peter Lemke, "The Response of Polar Sea Ice to Climate Variability and Change," in *Climate of the 21st Century: Changes and Risks*, ed. Jose L. Lozán et al. (Hamburg: GEO, 2001), pp. 206–11.

96 One fine summer morning: John Kelley is not sure this story really happened, at least in the way that Kenny Toovak tells it, but he likes it and he therefore encourages the retelling. Toovak, on the other hand, seemed to remember the story as a real event.

97 Elders across the Arctic: A volume that contains many studies of pan-Arctic indigenous knowledge about climate change has examples of the reduced predictability of the weather by elders: Igor Krupnik and Dyanna Jolly, eds., *The Earth Is Faster Now: Indigenous Observations of Arctic Environmental Change* (Fairbanks: Arctic Research Consortium of the United States, 2002).

97 Atmospheric scientists agreed: Jamie Morison of the University of Washington made this statement in an interview on August 8, 2001.

97 versions of a favorite local story: Not all the variants I have heard can be true, so I've chosen my favorite details from this modern legend.

98 Money allowed him to buy fuel oil: The Iñupiat traditionally lived in energy-efficient dugout sod houses kept warm enough by their seal oil lamps that people wore little clothing indoors. Missionaries arriving around the turn of the twentieth century for-

bade nudity and eventually moved villagers into frame houses. Driftwood that had been plentiful on Arctic beaches was soon exhausted in the effort to keep the drafty frame houses warm in the winter, so Eskimos resorted to seal oil and blubber for heat. Vilhjálmur Stefánsson covers much of this in *My Life with the Eskimo* (1913; repr., New York: Collier Books, 1962).

98 Some Alaska Natives responded to racism: A report issued in 2001 by the Alaska Governors Commission on Tolerance documented various examples of racial prejudice against Alaska Natives, but investigation wasn't really necessary to make the point. Alaska's majority white legislature and voters have in recent years passed several laws that Natives perceive as punitive, most blatantly a referendum making English the official language.

99 he was honored by scientific gatherings: See Maxwell E. Britton's foreword to *Fifty More Years Below Zero: Tributes and Meditations for the Naval Arctic Research Laboratory's First Half Century at Barrow, Alaska*, ed. David W. Norton (Fairbanks: Arctic Institute of North America and University of Alaska Press, 2001). See also chapter 10, pp. 282–84.

99 "I was scared of white people.": Karen Brewster, "Historical Perspectives on Iñupiat Contributions to Arctic Science at NARL," in *Fifty More Years*, p. 24.

99 three hundred Iñupiat worked there: Ibid., pp. 23–24.

100 most of the Arctic biological specimens: Max Brewer and John Schindler, "Introduction to Alaska's Original Naturalists," in *Fifty More Years*, pp. 9–10.

100 why didn't Kenny: John Hobbie, principal investigator of the Ecosystems Center at the Marine Biological Laboratory at Woods Hole, Massachusetts, made this point in an interview on February 18, 2003.

101 One such amendment: The act allowed Alaska Native corporations alone to sell "net operating losses" to other corporations, which could then deduct the losses from their tax liability. ASRC generated losses by writing down the value of land with diminished prospects of oil discovery, selling the losses to Dun and Bradstreet and Clorox for $90 million. Other Native corporations liquidated timber holdings at fire sale prices to generate paper losses they could sell before the loophole closed in 1988. Hal Bernton covered the story in the *Anchorage Daily News*: April 29, 1988; August 21, 1990; February 17, 1995; and other dates.

101 Oliver's possession of three-piece suits: Tom Kizzia, "Election of Chairman Reflects AFN Dilemma," *Anchorage Daily News*, October 26, 1985.

102 George and Maggie Ahmaogak fired off a letter: This account is based on interviews with Glenn Sheehan, Richard Glenn, Jackie Grebmeier, and Tom Pyle.

102 Grebmeier and Cooper findings: Lee W. Cooper, Jacqueline M. Grebmeier, et al., "Seasonal Variation in Sedimentation of Organic Materials in the St. Lawrence Island Polynya Region, Bering Sea," *Marine Ecology Progress Series* 226 (January 31, 2002): 13–26; and Jacqueline M. Grebmeier and Kenneth H. Dunton, "Benthic Processes in the Northern Bering/Chukchi Seas: Status and Global Change," *Impacts of Changes in Sea Ice and Other Environmental Parameters in the Arctic: Report of the Marine Mammal Commission Workshop, Girdwood, Alaska, 15–17 February, 2000*, ed.

Henry P. Huntington. Available from the Marine Mammal Commission, Bethesda, Maryland.

103 by far the largest lab of its kind: Gary A. Laursen, John J. Kelley, and Steven L. Stephenson, "Historical Perspectives on the Naval Arctic Research Laboratory, 1965 to 1980," in *Fifty More Years*, pp. 243–48.

104 unreliable dial-up connections: The connection speed was remedied in 2003.

105 candidates many considered mentally unstable: In 2002, Ted Stevens's Democratic opponent was Frank Vondesaar, who believed the senator was persecuting him by sending secret police to rearrange things in his house, and for that reason he did not have phone service. In 1996, Theresa Obermeyer was the Democratic nominee. She was motivated by an obsession that various powers, including Senator Stevens, had unjustly foiled her husband's attempts to achieve membership in the Alaska bar.

108 "You're the sub for the substitute.": I have chosen to omit the substitute teacher's name from this quotation.

5. THE ICE

112 his cousin Roy Ahmaogak: This Roy is two generations removed from the Roy Ahmaogak of Kenny Toovak's story in the last chapter.

112 functioned as cocaptains: Richard Glenn indicated *cocaptain* is not quite right, but it was as close as we could come in a single word.

114 the 1997 break-off: Craig George, Karen Brewster, et al., "Iñupiat Hunters and Shorefast Ice: How a Dynamic System Can Surprise Those Who Use It," submission to *Arctic*, in draft, describes the event and gives the figure of 154 stranded, which conflicts with some other accounts. Besides my own interviews, I also relied on two articles in the *Anchorage Daily News* by Rachel D'Oro and Don Hunter: "Rescuers Lift 142 Whalers Off Ice" (May 19, 1997) and "Drifting Island of Ice 'Was Quite Exciting' " (May 20, 1997).

116 Some are named right away: Jana Harcharek told me about the tradition of names and other Iñupiaq spiritual matters; more of that is covered in chapter 9.

117 one of Barrow's greatest whaling elders: Malik's English name, which I never heard used, was Ralph Ahkivgak.

118 a day or two after a big east wind can be a dangerous time: George, Brewster, et al., "Iñupiat Hunters and Shorefast Ice." Russell Page of the National Weather Service and Richard Glenn helped me understand the theory of how the cessation of the east wind could break off the ice.

120 an elder might be thirty years old: Lt. P. Henry Ray, *Report of the International Polar Expedition to Point Barrow, Alaska* (Washington: Government Printing Office, 1885), p. 44.

121 satellite pictures of the sea ice and the forecast: To see these products, go to arh.noaa.gov and click on Ice Desk.

124 Sea ice is an intrinsically fascinating: I owe much of my understanding of these processes to the patience of David Cole.

125 freshwater spreads over the top: Teneil Uttal et al., "Surface Heat Budget of the Arctic Ocean," *Bulletin of the American Meteorological Society* 83, no. 2 (February 2002).

125 Tiny organisms live: Hajo Eicken of the Geophysical Institute at the University of Alaska Fairbanks and Christopher Krembs of the University of Washington are studying these phenomena and showed me the slime-producing organisms in Eicken's lab on March 29, 2002.

125 Ice algae and tiny animals are food: Rolf Gradinger of the University of Alaska Institute for Marine Science, who was studying tiny ice organisms at BASC in April 2002, showed me samples and explained the process.

125 At one remarkable site: Lee W. Cooper, Jacqueline M. Grebmeier, et al., "Seasonal Variation in Sedimentation of Organic Materials in the St. Lawrence Island Polynya Region, Bering Sea," *Marine Ecology Progress Series* 226 (January 31, 2002).

125 a large creature like a sea worm: Dave Norton found the specimen and told me the story. The species is *Halicryptus higginsi.*

125 the theory of continental drift: Wegener's idea was not accepted until the 1960s.

126 remain too poorly understood: Lew Shapiro of the University of Alaska Fairbanks made this observation to me on May 6, 2002.

126 266 researchers and 900 tons of supplies: These figures come from Gary A. Laursen, John J. Kelly, and Steven L. Stephenson, "Historical Perpectives on the Naval Arctic Research Laboratory, 1965 to 1980," in *Fifty More Years Below Zero: Tributes and Meditations for the Naval Arctic Research Laboratory's First Half Century at Barrow, Alaska,* ed. David W. Norton (Fairbanks: Arctic Institute of North America and University of Alaska Press, 2001), p. 248. The USSR operated many more and larger ice camps than the United States, but the findings of their more active Arctic research program did not make it to the West; even today, those measurements and discoveries are not generally integrated into science outside Russia.

126 "It turned out essentially to be a flop": Nonetheless, AIDJEX was a success with many important discoveries, "the most successful scientific operation carried out in the Arctic Basin," in the view of W. F. Weeks "NARL and Research on Sea Ice and Lake Ice," in *Fifty More Years*, p. 180.

135 J. R. Leavitt's Crew: J. R. Leavitt is Oliver Leavitt's cousin, and David Leavitt is his uncle.

135 The whalers were on the seaward side of a long, wet crack: This account of the break-off is constructed from interviews with Richard Glenn, Patuk Glenn, Oliver Leavitt, Clayton Hopson, Billy Jens Leavitt, Brian Ahkiviana, Russell Page, and Doug Mealor, chief pilot for North Slope Borough Search and Rescue. I also drew on a May 15, 2002, article in the *Anchorage Daily News* by Anne Marie Tavella, "Ice Shatters, Sends Whalers Out into Arctic Ocean."

140 Gray whales were ample that night: This account is based largely on Joel Gay's reporting in the *Anchorage Daily News*, June 27 and July 9, 2002.

6. THE SUPERCOMPUTER

143 Charles Keeling, discoverer of the rapid increase: A readable account of this event is contained in William K. Stevens, *The Change in the Weather: People, Weather, and the Science of Climate* (New York: Random House, 1999), pp. 139–42.

143 The sun's energy: Some bizarre life forms obtain energy from geothermal vents.

144 the planet would average –18 degrees C: Ian Allison, glaciology program leader, Australian Antarctic Division, www.antdiv.gov.au.

144 those hundreds of millennia: P. Falkowski et al., "The Global Carbon Cycle: A Test of Our Knowledge of Earth as a System," *Science*, vol. 290, no. 5490 (October 13, 2000): 291–96.

145 Current levels of human energy use: Carbon from natural respiration and decay amounts to approximately 100 billion metric tons annually, while carbon from fossil fuels and human land use is about 8 billion metric tons annually. This figure differs from the human use of total photosynthesis products, estimated at 10 to 55 percent (see chapter 10), because that number is based on terrestrial net primary production on land, which does not include the photosynthesis necessary for the plants' own respiration or life in the oceans.

145 A human's daily diet of 2,000 calories: I use the familiar meaning of the word *calorie*; the correct word would be *kilocalorie*, or 1,000 calories. Also, I have rounded figures throughout this discussion.

145 Wood is half carbon: Energy and carbon conversion figures are widely available; I made these calculations based on figures found on a Department of Energy Web site, bioenergy.ornl.gov/papers/misc/energy_conv.html.

145 1,750 pounds of wood growth: Based on .647 pounds of carbon dioxide per passenger mile (1999) for a 5,000-mile trip, converted to carbon equivalent.

145 a ton of carbon per acre per year: The figures in this sentence were provided by Steve Pacala of Princeton University.

145 That's far more forest than we have: Forested area of the United States is roughly 750 million acres. However, I am presenting these figures only for the sake of explanation. The carbon uptake of the oceans is also important.

146 atmospheric CO_2 was 288 parts per million: Carbon Dioxide Information Analysis Center (CDIAC), U.S. Department of Energy (cdiac.esd.ornl.gov); Stevens, *Change in the Weather*; NOAA CMDL Summary Report 26.

146 the amount of carbon stored in the forests: This and other forestry data are from *The State of the Nation's Ecosystems: Measuring the Lands, Waters, and Living Resources of the United States* (Washington: H. John Heinz III Center for Science, Economics, and the Environment, 2002).

146 ocean water dissolved about 2 billion metric tons of carbon: Ben I. McNeil, Richard J. Matear, et al., "Anthropogenic CO_2 Uptake by the Ocean Based on the Global Chlorofluorocarbon Data Set," *Science*, vol. 299, no. 5608 (January 10, 2003): 235–39.

146 that drop could be severe: Falkowski et al., "Global Carbon Cycle"; Jorge L. Sarmiento and Corinne Le Quéré, "Oceanic Carbon Dioxide Uptake in a Model of Century-Scale Global Warming," *Science*, vol. 274, no. 5291 (November 22, 1996): 1346–50; Richard B. Rivkin and Louis Legendre, "Biogenic Carbon Cycling in the Upper Ocean: Effects of Microbial Respiration," *Science*, vol. 291, no. 5512 (March 23, 2001): 2398–400.

147 more than humans have ever released: The 450 billion metric ton figure is from an

interview with Walter Oechel on October 11, 2001; total human carbon emissions from fossil fuels and land use changes total 408 billion metric tons and total atmospheric carbon 780 billion, from the CDIAC.

147 The rising trend of carbon dioxide: The data in this paragraph are from CMDL Summary Report 26.

149 we don't know enough to make a prediction: This is based on the Oechel interview and on Oechel et al., "Acclimation of Ecosystem CO_2 Exchange in the Alaskan Arctic in Response to Decadal Climate Warming," *Nature*, vol. 406, no. 6799 (August 31, 2000): 978–81.

149 That asterisk next to determinism Lorenz called chaos: Edward N. Lorenz, *The Essence of Chaos* (Seattle: University of Washington Press, 1993), pp. 6–15.

149 Even without understanding nonlinear mathematics: Ibid., pp. 102–10.

150 Weather services extend the range: European Center for Mid-Range Weather Forecasts, www.ecmwf.int.

151 a working climate out of pure mathematics: Syukuro Manabe and Robert F. Strickler, "Thermal Equilibrium of the Atmosphere with a Convective Adjustment," *Journal of the Atmospheric Sciences* 21, no. 4 (July 1964): 361–85. These calculations had been attempted since the beginning of the century; by introducing convection, Manabe succeeded where others failed.

151 A doubling of CO_2 from preindustrial levels: Syukuro Manabe and Richard T. Wetherald, "Thermal Equilibrium of the Atmosphere with a Given Distribution of Relative Humidity," *Journal of the Atmospheric Sciences*, vol. 24, no. 3 (May 1967): 241–59.

151 We met in a little office: Interview with Manabe, October 15, 2002.

151 In 1975, Suki published a model in three dimensions: Syukuro Manabe and Richard T. Wetherald, "The Effects of Doubling the CO_2 Concentration on the Climate of a General Circulation Model," *Journal of the Atmospheric Sciences*, vol. 32, no. 1 (January 1975): 3–15.

152 Climate modelers started at the beginning: Warren M. Washington and Claire L. Parkinson, *An Introduction to Three-Dimensional Climate Modeling* (Mill Valley: University Science Books, 1986), pp. 57–59.

152 A full working model needed: B. A. Boville, "Toward a Complete Model of the Climate System," in *Numerical Modeling of the Global Atmosphere in the Climate System*, eds. Philip Mote and Alan O'Neill (Boston: Kluwer Academic Publishers, 2000), pp. 419–21.

153 Field studies often focused: This is my own observation based on time spent with various field projects.

154 "Q: You are sure": Isaac Asimov, *Foundation* (1951; repr. New York: Avon, 1966), pp. 25–26.

154 Flux adjustments: Boville, "Toward a Complete Model," pp. 424–25.

155 Parameterizations dealt with these unknowns either by: A. Henderson-Sellers and K. McGuffie, *A Climate Modelling Primer* (Chichester: John Wiley & Sons, 1987), pp. 49–52.

155 GCMs' treatment of sea ice: Boville, "Toward a Complete Model," pp. 429–30; Drew Shindell, "Whither Arctic Climate," *Science*, vol. 299, no. 5604 (January 10, 2003): 215–16.

155 Some showed ice disappearing entirely: Cecilia Bitz, Polar Science Center, University of Washington, lecture at the International Arctic Research Center, Fairbanks, March 27, 2002.

155 GCMs had trouble showing the buildup of clouds: This paragraph is based on an interview with John Walsh at the University of Alaska Fairbanks on March 28, 2002.

156 Department of Energy's Atmospheric Radiation Measurement (ARM) Web site: www.arm.gov.

158 The fastest computers would need to be ten times faster: Walsh interview, August 6, 2001.

159 *Science, Nature*, and the *Washington Post* all ridiculed the project: Paul Selvin, "Alaskan Pork: Aurora Fantasia," *Science*, vol. 250, no. 4984 (November 23, 1990): 1073; Christopher Anderson, "Harnessing Northern Lights," *Nature*, vol. 348, no. 6297 (November 8, 1990): 101; Susan Cohen, "Pork in the Sky," *Washington Post*, November 10, 1991.

159 In 2002, the center installed: John Markoff, "Japanese Supercomputer Finds a Home in Alaska," *New York Times*, June 14, 2002.

159 A variety of computers: ARSC research liaison Guy Robinson told me about the computers and gave Robin and me our tour on July 1, 2002.

160 "I am perhaps not a normal scientist": Syun-Ichi Akasofu, *Exploring the Secrets of the Aurora* (Boston: Kluwer Academic Publishers, 2002) pp. 207, xviii. Akasofu's ideas about the progress of science draw on Thomas Kuhn, *The Structure of Scientific Revolutions*, 3rd ed. (Chicago: University of Chicago Press, 1996), but Akasofu approaches the topic as a practitioner. I have also drawn on my own profile of him, published in the *Anchorage Daily News*, October 19, 1997.

160 the IARC opening: This history of IARC is based on my interviews, with dates from the IARC Web site, at www.iarc.uaf.edu.

160 The Senate Appropriations Committee formally urged: Senator Ted Stevens, press release, January 24, 2003.

165 ten years and $150 million away: W. Wayt Gibbs, "Cybernetic Cells," *Scientific American* 285, no. 2 (August 2001), pp. 54–57.

165 One such idea under study: Ralph Lorenz, "Full Steam Ahead—Probably," *Science*, vol. 299, no. 5608 (February 7, 2003): 837.

165 a would-be Kepler: My description, which Johnny Lin eschews.

166 "It's probably best understood": Lin was careful to note that his application of these theories to weather and climate is not all original. He said information theory was first applied to precipitation in 1993 by A. A. Tsonis and J. B. Elsner.

167 "two things they don't understand": Roe attributes this line originally to Phillip England of Oxford University.

168 Ice cores drilled in Greenland: Paul Andrew Mayewski and Frank White, *The Ice Chronicles: The Quest to Understand Global Climate Change* (Hanover: University Press of New England, 2002), pp. 91–95.

168 reasonable doubt: This was Bernie Zak's formulation of the level of certainty we have now.

168 the largest factor that was changing: From 1850 to 1992, the change in radiative forcing from greenhouse gases was at least double the next strongest effect, and probably much larger: G. J. Boer, "Simulating Future Climate," in Mote and O'Neill, *Numerical Modeling*, p. 491.

170 Suki and Ron built a model in the 1980s: S. Manabe and R. J. Stouffer, "Two Stable Equilibria of a Coupled Ocean-Atmosphere Model," *Journal of Climate* 1, no. 9 (September 1988): 841–66.

170 The THC has varied in strength: Peter U. Clark et al., "The Role of the Thermohaline Circulation in Abrupt Climate Change," *Nature*, vol. 415, no. 6874 (February 21, 2002) 863–69.

170 Preliminary measurements suggested: Richard A. Kerr, "Another Way to Take the Ocean's Pulse," *Science*, vol. 299, no. 5605 (January 17, 2003): 337.

7. THE SIGNS

171 A long article in *Outside*: Jack Hitt, "If I Can Take It There, I Can Take It Anywhere," *Outside* 21, no. 6 (June 1996), p. 84. Hitt's piece is not unique; he joins a long history of journalists projecting their fantasies on the Iñupiat. Among the most bizarre cases is Tom Rose's, *Freeing the Whales: How the Media Created the World's Greatest Non-Event* (New York: Birch Lane Press, 1989), which contains this imaginative passage on p. 50: "For a town that had been using whale meat as its primary means of exchange, Barrow's sudden wealth brought enormous change. Not only were Barrow's elders ignorant about managing money, most of them had never seen it. Illiterate subsistence whalers who didn't speak a word of English were suddenly overseeing multi-billion dollar portfolios from the earthen floors of their sod huts." Ironically, the book is a work of media criticism.

173 not in proportion with oddities of the earth's path: Richard A. Muller and Gordon J. MacDonald, "Glacial Cycles and Astronomical Forcing," *Science*, vol. 277, no. 5323 (July 11, 1997): 215–18.

173 first found the trace of a sixty-to-eighty-year cycle: I. V. Polyakov and M. A. Johnson, "Arctic Decadal and Interdecadal Variability," *Geophysical Research Letters* 27, no. 24 (2000): 4097.

173 it erased the Arctic amplification: Igor Polyakov et al., "Observationally Based Assessment of Polar Amplification of Global Warming," *Geophysical Research Letters* 29, no. 18, (2002): 1878, and interviews with Polyakov. He disagrees, however, that his finding could reduce the justification for funding Arctic research.

174 When the categories were tabulated: N. A. Marchenko, R. L. Colony, V. A. Nizovtcev, "Extreme Natural Events in Central Russia Over the Last Millennium," *Changes in the Atmosphere-Land-Sea System in the Amerasian Arctic: Proceedings of the Arctic Regional Centre* 3 (Vladivostok: Dalnauka, 2001), pp. 23–28.

174 Arctic Oscillation: John M. Wallace and David W. J. Thompson, "Annular Modes and Climate Prediction," *Physics Today* 55, no. 2 (February 2002): 28; David W. J. Thomp-

son and John M. Wallace, "Regional Climate Impacts of the Northern Hemisphere Annular Mode," *Science*, vol. 293, no. 5527 (July 6, 2001): 85.

176 AO remained a controversial hypothesis: The controversy concerns the relationship of the AO to the North Atlantic Oscillation (NAO), which was discovered in 1932. Critics say the part of the AO that works is simply a restatement of the NAO, whereas the balance of the pattern is not physically consistent. See Maarten H. P. Ambaum, Brian J. Hoskins, and David B. Stephenson, "Arctic Oscillation or North Atlantic Oscillation," *Journal of Climate*, vol. 14, no. 16 (August 15, 2001): 3495–507. Wallace counters that the NAO is part of the AO that extends over the Atlantic due to the configuration of the continents. As evidence, he shows that the Antarctic has a pattern like the Arctic Oscillation that is more symmetrical due to the absence of north-south-oriented land masses.

176 the paper that first announced their finding: David W. J. Thompson and John M. Wallace, "The Arctic Oscillation Signature in the Wintertime Geopotential Height and Temperature Fields," *Geophysical Research Letters*, vol. 25, no. 9 (May 1, 1998): 1297–300.

177 University of Washington scientists called for fishery managers: Nathan J. Mantua et al., "A Pacific Interdecadal Climate Oscillation with Impacts on Salmon Production," *Bulletin of the American Meteorological Society*, vol. 78, no. 6 (June 1997): 1069–79.

177 the authors of a paper in *Science* suggested: Francisco P. Chavez et al., "From Anchovies to Sardines and Back: Multidecadal Change in the Pacific Ocean," *Science*, vol. 299, no. 5604 (January 10, 2003): 217–21.

177 That idea could help explain: Nicholas J. Shackleton, "The 100,000-Year Ice-Age Cycle Identified and Found to Lag Temperature, Carbon Dioxide, and Orbital Eccentricity," *Science*, vol. 289, no. 5486 (September 15, 2000): 1897–902.

177 British scientists found: Arnold H. Taylor, J. Icarus Allen, Paul A. Clark, "Extraction of a Weak Climatic Signal by an Ecosystem," *Nature*, vol. 416, no. 6881 (April 11, 2002): 629.

179 Alaska's glaciers alone were producing enough water: This paragraph is based on interviews with Keith Echelmeyer, August 23, 2001 and other dates; Anthony Arendt, Keith Echelmeyer, et al., "Rapid Wastage of Alaska Glaciers and Their Contribution to Rising Sea Level," *Science*, vol. 297, no. 5580 (July 19, 2002): 382–86.

179 the eighty-four-year record of a statewide betting pool: Raphael Sagarin and Fiorenza Micheli, "Climate Change in Nontraditional Data Sets," *Science*, vol. 294, no. 5543 (October 26, 2001): 811.

180 authors hedged and called for more research: M. C. Serreze et al., "Observational Evidence of Recent Change in the Northern High-Latitude Environment," *Climatic Change* 46, no. 1–2 (July, 2000): 159–207; J. Overpeck et al., "Arctic Environmental Change of the Last Four Centuries," *Science*, vol. 278, no. 5341 (November 14, 1997): 1251–56.

180 elders would tell stories: These examples are all from *Puiguitkaat: The 1978 Elders'*

Conference (Barrow: North Slope Borough Commission on History and Culture), pp. 402, 408, 55.

183 the endings were compressing: This was the observation of Edna MacLean, editor of the definitive Iñupiaq dictionary.

185 Still, traditional knowledge gained a reputation: Skeptics also observe that traditional knowledge almost always supports Natives' political positions regarding resource allocation and environmental protection. That criticism strikes me as unfounded on two counts. First, it is a circular argument: if the Natives base their political views on their traditional knowledge, then the two would be expected to be in concert. This point is not contradicted by the fact that Native communities more often call for greater, not smaller, harvests of wildlife; they would have no reason to call for lower limits on themselves when they could self-impose such limits. Second, there is a major counterexample to this theory. Iñupiat observations that climate change is real do not support their political advocacy for increased oil and gas development.

186 "One result of the symposium": Henry P. Huntington, Harry Brower, Jr., and David W. Norton, "The Barrow Symposium on Sea Ice, 2000: Evaluation of One Means of Exchanging Information between Subsistence Whalers and Scientists," *Arctic* 54, no. 2 (June 2001): 201–06. I have spelled out acronyms in the original text.

188 "Researchers may have contaminated the knowledge": Anonymous comments on NSF Arctic System Science Program proposal 0125141, Dave Norton, principal investigator, "Collaborative Research: A Synthesis Approach to Link Glaciological and Remote-Sensing Studies with Natural History and Traditional Knowledge, by Case Studies of Coastal Ice."

189 "If thought is like the keyboard of a piano": Virginia Woolf, *To the Lighthouse* (New York: Harcourt Brace Jovanovich, 1927), pp. 53–54.

189 Intuition is real: These and many other examples are contained in David G. Myers's fascinating book *Intuition: Its Powers and Perils* (New Haven: Yale University Press, 2002). I am indebted to Myers, who also assisted in personal communication.

191 Advanced magnetic brain imaging shows how this works: Ognjen Amidzic et al., "Pattern of Focal Gamma Bursts in Chess Players," *Nature,* vol. 412, no. 6847 (August 9, 2001): 603.

192 IBM's Deep Blue: Bruce Weber, "What Deep Blue Learned in Chess School," *New York Times,* May 18, 1997.

194 *Science* published an article: Carl Wunsch, "What Is the Thermohaline Circulation?" *Science,* vol. 298, no. 5596 (November 8, 2002): 1179–81.

194 He also wrote a funny piece: Norbert Untersteiner, "Cite This Letter!" *Physics Today,* 48, no. 4 (April 1995).

194 Later, researchers published a real paper: Tom Clark, "Copied Citations Give Impact Factors a Boost," *Nature,* vol. 423, no. 6938 (May 22, 2003): 373.

195 I met researchers in Barrow: I have not identified these researchers because a fair treatment of their project would require more space than I think is justified.

196 a newsletter intended to help make it all clear: "International Collaboration in the Paleosciences: The Beringian Connections," *Witness the Arctic* 8, no. 2 (Winter 2000–01), published by the Arctic Research Consortium of the United States, Fairbanks.

197 HARC, PARCS, ARC, ARCUS, ARCSS, and ARCMIP: Human Dimensions of the Arctic System, Paleoenvironmental Arctic Sciences, U.S. Arctic Research Commission, Arctic Research Consortium of the United States, Arctic System Science, and Arctic Regional Model Intercomparison Project.

197 Arctic sea ice had thinned by 42 percent: D. A. Rothrock, Y. Yu, and G. A. Maykut, "Thinning of the Arctic Sea-Ice Cover," *Geophysical Research Letters*, vol. 26, no. 3 (December 1, 1999): 3469–72.

198 a name that would catch on: This account of the naming of Unaami came from an interview with Morison on August 8, 2001.

8. THE CAMPS

203 the boost was only temporary: Gaius R. Shaver et al., "Global Change and the Carbon Balance of Arctic Ecosystems," *Bioscience* 42, no. 6 (June 1992): 433.

207 Matthew's key shrub paper: Matthew Sturm et al., "Snow-Shrub Interactions in Arctic Tundra: A Hypothesis with Climatic Implications," *Journal of Climate*, vol. 14, no. 3 (February 1, 2001): 336.

207 Matthew's style of capturing patterns: This is my observation, not Matthew Sturm's.

208 Birch happened to be especially efficient: Gaius R. Shaver et al., "Species Composition Interacts with Fertilizer to Control Long-Term Change in Tundra Productivity," *Ecology* 82, no. 1 (2001): 3163.

209 A modeling project led by Terry Chapin: F. Stuart Chapin et al., "Summer Differences among Arctic Ecosystems in Regional Climate Forcing," *Journal of Climate*, vol. 13, no. 12 (June 15, 2000): 2002.

211 Dirk: Since this student may not have been aware I was a journalist, I have given him a fictitious name.

212 he was angry: Ted Stevens, press release, March 4, 2003.

212 The authors discounted: National Academy of Sciences, Board on Environmental Studies and Toxicology and Polar Research Board, *Cumulative Environmental Effects of Oil and Gas Activities on Alaska's North Slope* (Washington: National Academies Press, 2003).

212 "The statements about visual effects": This passage is from an e-mail that Richard Glenn submitted to the authors of the report on November 11, 2002.

213 The land that became parks in the Lower 48: Sometimes, national parkland was easy to set aside specifically because it lacked biological value to people or animals. The big fights came with attempts to protect productive land. The story of Grand Teton National Park exemplifies this point; it is well told in Robert W. Righter, *Crucible for Conservation: The Struggle for Grand Teton National Park* (Boulder: Colorado Associated University Press, 1982). The entire story of the National Park Service's evolution from valuing aesthetics and recreation to valuing natural qualities is told in

Richard West Sellars, *Preserving Nature in the National Parks* (New Haven: Yale University Press, 1997).

219 Henry Adams: Henry Adams, *The Education of Henry Adams* (1907; repr., New York: Modern Library, 1931), pp. 72–73.

222 "For this new creation": Ibid., pp. 496–97.

225 The NSF was spending more than three times as much: *National Science Foundation FY 2004 Budget Request to Congress*, p. 321. U.S. Polar Research Programs totaled $262 million, including $29 million for Arctic Research Support and Logistics and $147 million for Antarctic Operations and Science Support. The actual research in each region (science rather than support funding) was $41 million for the Arctic and $44 million for the Antarctic.

226 2.5 percent of federal spending: I developed these figures based on estimated federal spending for fiscal year 2004 of $2,229 billion from the Office of Management and Budget, *Budget for the Fiscal Year 2004*, p. 115; total civilian R&D spending for fiscal year 2004 of $55.8 billion from *Science*, vol. 299, no. 5608 (February 7, 2003): 808; total NSF spending of $5.5 billion and total NSF spending on the Arctic of $70 million from the *National Science Foundation FY 2004 Budget Request to Congress*, pp. 1 and 321.

9. THE SPIRIT

231 "unnecessary and cruel": This quote is from Richard Lloyd Parry, "Whaling Ban to Remain in Force after Summit Blocks Iceland," *The Independent*, May 21, 2002. I have also drawn on Parry, "Japan Fights to End 16-Year Whaling Ban," *The Independent*, May 20, 2002; Tom Kizzia, "Inupiat Lose Bowhead Appeal," *Anchorage Daily News*, May 24, 2002; and James Brooke, "An Environmentalist Who Loves to Eat Whales," *New York Times*, October 19, 2002; as well as interviews with those mentioned in the text.

233 three levels of a right to life: Peter Singer, *Practical Ethics* (Cambridge: Cambridge University Press, 1979). The list of potentially self-conscious animals is on p. 103. Singer excused Eskimos for killing animals to survive, but only as a necessity, and he particularly condemned killing whales (pp. 55 and 98, respectively).

233 PETA advertising campaign: Michelle Morgante, "Jewish Groups Decry PETA's Holocaust Ads," Associated Press, March 1, 2003.

233 There is no uniquely human gene: Shirley Tilghman lecture at Princeton University, October 14, 2002; research announced after her lecture reduced the similarity of human beings and chimpanzees from 98.7 percent to 94–95 percent (Helen Pearson, "Chimps Expose Humanness," *Nature Science Update* [www.nature.com/nsu], April 29, 2003).

234 *Wilderness & the American Mind*: Roderick Frazier Nash, *Wilderness & the American Mind*, 4th ed. (New Haven: Yale University Press, 2001).

235 an essential new element: This is a slight caricature, but Muir frequently recounted his spiritual elevation while exposed to the elements, and he traveled rough, like a modern backpacker. He wrote that "only by going alone in silence, without baggage, can one truly get into the heart of the wilderness" (Nash, *Wilderness*, p. 282).

235 National Park Service: Richard West Sellars, *Preserving Nature in the National Parks* (New Haven: Yale University Press, 1997), covers the history of change within the park service. I traveled to many of the parks myself and observed and read about their practices while writing a travel book, *Family Vacations in the National Parks*, 2nd ed. (New York: Wiley, 2002).

236 One strain of thinking: Nash, *Wilderness*, pp. 245–46.

236 drilling would be a "desecration": Lieberman was speaking on the Senate floor on March 18, 2003.

237 An Audubon Society official said: Stan Senner, executive director of the Alaska branch of the Audubon Society, quoted in Zev Chafets, "Holy Ground: Spiritualism, Not Logic, Drives Environmentalist Objections to Oil Drilling in Alaska," New York *Daily News*, June 3, 2001.

237 They turned to philosophy: Nash, *Wilderness*, pp. 238–39.

237 strip mining raped land literally: Roderick Nash, "Do Rocks Have Rights?" in *Small Comforts for Hard Times: Humanists on Public Policy*, eds. Michael Mooney and Florian Stuber (New York: Columbia University Press, 1977), pp. 120–34. Nash further develops these thoughts, but in a guise of disinterest, in *The Rights of Nature: A History of Environmental Ethics* (Madison: University of Wisconsin Press, 1989). Readers familiar with his work may argue that we reach the same conclusion, since he indicates that a broadening respect for nature mirrors indigenous peoples' original view of the environment. But Nash appropriates that idea for an entirely different purpose: Alaska Natives' conception of membership in nature is the reverse of his aim of preserving it separate from mankind.

237 "Keeping human hands off": Nash, *Wilderness*, p. 388.

237 "the intrinsic rights": Ibid., p. 389.

237 the dawning of a person's unique right to life: Singer, *Practical Ethics*, p. 124.

238 consciousness is essentially a metaphor: Julian Jaynes, *The Origin of Consciousness in the Breakdown of the Bicameral Mind* (Boston: Houghton Mifflin, 1976), pp. 21–66.

238 *Hamlet* demonstrates the pitfall of self-conscious thought: I am again indebted to David G. Myers, *Intuition: Its Powers and Perils* (New Haven: Yale University Press, 2002).

239 Lutz and Sitka spruce: Leslie A. Vierick and Elbert L. Little Jr., *Alaska Trees and Shrubs* (1972; repr., Fairbanks: University of Alaska Press, 1986); and Ed Berg, personal communication.

239 Ed's next-door neighbors: John and Mary Jane Shows.

240 Entomologists thought Kachemak Bay was immune: This paragraph is based on interviews with Ed Berg and Edward Holsten, an entomologist with the Forest Service in Anchorage; and Edward Holsten et al., *Insects and Diseases of Alaskan Forests* (Anchorage: USDA Forest Service Alaska Region, 2001).

240 Ed turned to low-tech tree ring study: At this writing, Berg's work was unpublished.

240 Four million acres of spruce died: The four-million-acre figure and the course of the outbreak are covered in USDA Forest Service and Alaska Department of Natural Resources, *Forest Insect and Disease Conditions in Alaska—2001: General Technical Re-*

port R10-TP-102 (Anchorage, 2002). That this kill is the largest ever is from National Assessment Synthesis Team, U.S. Global Change Research Program, *Climate Change Impacts on the United States: The Potential Consequences of Climate Variability and Change*; *Overview* (Cambridge: Cambridge University Press, 2000), p. 77.

242 two million acres scorched: Alaska Department of Natural Resources, Division of Forestry; annual fire statistics are found at www.dnr.state.ak.us/forestry.

242 measurements were just beginning: In 2002, Jim Randerson of Caltech was doing this work near Delta Junction with towers like those used by Walt Oechel in Barrow. Randerson's funding came from seed money provided by Caltech when he set up his lab there.

242 an Anchorage snow plow operator: His name is David Wolfe.

242 Natives gathered their own observations: These observations come from a meeting I attended in Anchorage on August 20, 2002, and from the *Alaska Traditional Knowledge and Native Foods Database*, a project of the Alaska Native Science Commission and the Institute for Social and Economic Research at the University of Alaska Anchorage, at www.nativeknowledge.org.

244 species ranges shifted 6.1 kilometers: Camille Parmesan and Gary Yohe, "A Globally Coherent Fingerprint of Climate Change Impacts across Natural Systems," *Nature*, vol. 421, no. 6918 (January 2, 2003): 37.

247 Catholic philosopher Stanley Jaki: Stanley L. Jaki, *The Road of Science and the Ways to God* (Chicago: University of Chicago Press, 1978), p. vii.

247 "Today the same insults": Ibid., p. 15.

247 The Roman Catholic Church formally accepted: *The Cambridge Encyclopedia*, 3rd ed., ed. David Crystal (Cambridge: Cambridge University Press, 1997), p. 437.

247 Jaki pointed out: Jaki, *Road of Science*, pp. 40–49.

249 A great hunter and his wife lived alone: I have adapted a version told by Sagdluaq, a man of the Colville River, to Knud Rasmussen in 1924. Rasmussen traveled by dogsled from Greenland to Nome from 1921 to 1924, collecting stories that he translated from Iñupiaq and Inuit languages to Danish and then to English, and his writings are an excellent source for this material and for some cultural practices. The Kivgiq story is contained in both of the following books with slightly different translations (the latter of these volumes is also a good read): H. Ostermann, *The Alaskan Eskimos, as Described in the Posthumous Notes of Knud Rasmussen*, ed. E. Holtved, trans. W. E. Calvert (1952; repr., New York: AMS Press, 1976), pp. 38–42; and Knud Rasmussen, *Across Arctic America: Narrative of the Fifth Thule Expedition* (1927; repr., New York: Greenwood Press, 1969), pp. 323–27.

250 Kivgiq: Ostermann, *Alaskan Eskimos*, pp. 103–12. I have also relied on a document that explains the traditional and modern Kivgiq, Chris B. Wooley and Rex A. Okakok, "Kivgiq: A Celebration of Who We Are," presented to the 16th Annual Meeting of the Alaska Anthropological Association, March 3–4, 1989, Anchorage.

251 Their shamans communicated: Laurie Kingik, in *Puiguitkaat: The 1978 Elders' Conference* (Barrow: North Slope Borough Commission on History and Culture, 1981), p. 52.

251 the Iñupiat did not conceive: James L. Cox, *The Impact of Christian Missions on Indigenous Cultures: The "Real People" and the Unreal Gospels* (Lewiston: Edwin Mellen Press, 1991), pp. 11–13.

251 In one story: Elijah Kakinya, in *Puiguitkaat*, pp. 62–64.

251 Presbyterian minister Sheldon Jackson: These paragraphs are based largely on Cox, *Impact*, pp. 1–28. That the Iñupiat desperately needed health and education services, and that the missionaries relieved great physical suffering, is indisputable. The debate between assimilation or sovereignty, which dates at least to President George Washington's administration, is still going on.

252 Lela Kiana Oman: Oman's books include the riveting tale *The Epic of Qayak: The Longest Story Ever Told by My People* (Ottawa: Carlton University Press, 1995).

252 Vilhjálmur Stefánsson, traveling in 1908: Vilhjálmur Stefánsson, *My Life with the Eskimo* (1913; repr., New York: Collier Books, 1962), pp. 44, 88, 95–98.

253 dead relatives would return: Elijah Kakinya, in *Puiguitkaat*, pp. 78–79.

254 Dr. Simpson's 1855 monograph: *The Journal of Rochfort Maguire, 1852–1854*, vol. 2, ed. John Bockstoce (London: Hakluyt Society, 1988), pp. 549–50.

254 the Iñupiat still believed: This is my summary of statements and feelings expressed by various people at different times; it certainly does not represent the entire community or even, as far as I know, any one person.

255 "We were put on this earth": Comments at the Interior Regional Meeting, March 27, 1999, in *Alaska Traditional Knowledge and Native Foods Database*, established and maintained by the Alaska Native Science Commission and Institute for Social and Economic Research, University of Alaska Anchorage, at www.nativeknowledge.org.

10. THE CHALLENGE

256 Whalers from Utqiagvik: Murdoch, *Ethnological Results of the Barrow Expedition* (1892) (Washington: Smithsonian Institution Press, 1988), p. 54.

257 revive fall whaling: Fall whaling also occurred prior to the 1970s moratorium.

258 A man was roughly normal-sized prey: Most of this information on polar bears is from an interview with Tom Albert on September 10, 2001, before he retired from North Slope Borough Wildlife.

258 In the village of Point Lay: This is based on two articles I wrote in the *Anchorage Daily News*, December 10 and 12, 1990.

259 researchers found the bears growing skinnier: Ian Stirling et al., "Long-Term Trends in the Population Ecology of Polar Bears in Western Hudson Bay in Relation to Climate Change," *Arctic* 52, no. 3 (September 1999): 294–306.

260 as much as 1,500 feet: National Assessment Synthesis Team, U.S. Global Change Research Program, *Climate Change Impacts on the United States: The Potential Consequences of Climate Variability and Change; Foundation* (Cambridge: Cambridge University Press, 2001), p. 293.

262 the lowest in fifty years: NOAA National Snow and Ice Data Center, "Arctic Sea Ice Shrinking, Greenland Ice Sheet Melting, According to Study" (press release at the

American Geophysical Union annual meeting, December 7, 2002; http://nsidc.org/news/press/20021207_seaice.html).

263 If the ice ever disappears entirely: Predictions on when this might happen are of little value; they vary from fifty years to well over a century.

263 Maritime experts expected: Richard A. Kerr, "A Warmer Arctic Means Change for All," *Science*, vol. 297, no. 5586 (August 30, 2002): 1490.

264 When Lieutenant Ray's expedition arrived: Murdoch, *Ethnological Results*, p. 53.

267 Ecosystems are at most 1 or 2 percent efficient: This point and the other material about organic and artificial photosynthesis in the paragraph are based on an interview February 28, 2003, with Nathan Lewis of Caltech and his article "Artificial Photosynthesis," *American Scientist* 83, no. 6 (November–December 1995): 534.

267 Human beings were already using: Stuart Rojstaczer, Shannon M. Sterling, and Nathan J. Moore, "Human Appropriation of Photosynthesis Products," *Science*, vol. 294, no. 5575 (December 21, 2001): 2549. Their estimate at a 95 percent confidence level ranges from 10 to 55 percent; however, even at the low end of the range the point holds that organic photosynthesis cannot supply most of human fuel needs. Others feel the number is more certain. The figure is based on terrestrial net primary production and does not include the photosynthesis that plants require to grow.

268 $40-a-gallon gasoline: Nathan Lewis's estimate.

268 an area about the size of Colorado: Based on global energy use of 12 TW. I have relied in part on a review article containing a summary of these technology issues: Martin I. Hoffert et al., "Advanced Technology Paths to Global Climate Stability: Energy for a Greenhouse Planet," *Science*, vol. 298, no. 5595 (November 1, 2002): 981.

268 a project to build a living cell: Carl Zimmer, "Tinker, Tailor: Can Venter Stitch Together a Genome from Scratch?" *Science*, vol. 299, no. 5609 (February 14, 2003): 1006.

268 fifty years or more away: I have omitted various more immediate options that could contribute to a solution, such as nuclear fission and tidal energy: all have environmental problems or limited application or both. With 85 percent of world energy needs currently supplied by fossil fuels, any replacement must be substantial in scale and have manageable environmental consequences.

268 we would still be on track: Hoffert et al., "Advanced Technology."

269 what to do with the leftover carbon: For a technical summary of options, see Klaus S. Lackner, "A Guide to CO_2 Sequestration," *Science*, vol. 300, no. 5626 (June 13, 2003): 1677–78.

269 A test marketplace: The Web site is www.chicagoclimatex.com.

270 "the same trap as the tobacco companies": Ross Gelbspan, *The Heat Is On: The High Stakes Battle over Earth's Threatened Climate* (Reading, Mass.: Addison-Wesley, 1997), pp. 79–86.

270 addressing carbon emissions saved the company money: Press conference, Anchorage, June 28, 2002.

270 a $225 million carbon mitigation study: Andrew Revkin, "Exxon-Led Group Is Giving a Climate Grant to Stanford," *New York Times*, November 21, 2002.

270 Bush followed the same path: White House press release, "Hydrogen Fuel: A Clean and Secure Energy Future," January 28, 2003.

270 Bush and his Texas oil allies still denied: In April 2003, ExxonMobil's Web site said, "Scientific research to improve understanding of climate change and its potential risk must continue, as there are still many uncertainties. We have invested over twenty years in climate research and will continue to be an active contributor to scientific understanding." On June 13, 2001, President Bush made these comments in the Rose Garden, posted in April 2003 on whitehouse.gov as representative of the administration's climate change policy: "We do not know how much our climate could, or will change in the future. We do not know how fast change will occur, or even how some of our actions could impact it. For example, our useful efforts to reduce sulfur emissions may have actually increased warming, because sulfate particles reflect sunlight, bouncing it back into space. And, finally, no one can say with any certainty what constitutes a dangerous level of warming, and therefore what level must be avoided."

270 the greenhouse effect was real: Soon after taking office, the Bush administration asked for the National Research Council to respond to a list of questions apparently designed to highlight the uncertainties in climate science, allowing less than a month to respond. The report that came back began with these sentences: "Greenhouse gases are accumulating in Earth's atmosphere as a result of human activities, causing surface air temperatures and subsurface ocean temperatures to rise. Temperatures are, in fact, rising." Committee on the Science of Climate Change, Division on Earth and Life Studies, National Research Council, *Climate Change Science: An Analysis of Some Key Questions* (Washington: National Academies Press, 2001).

270 doubters themselves were already feeling the change: oil exploration, Alaska Department of Natural Resources; trans-Alaska oil pipeline, Gunter Weller, University of Alaska Center for Global Change, March 25, 2002; missile defense, U.S. Arctic Resource Commission presentations at U.S. Army Corps of Engineers Cold Regions Research and Engineering Lab, October 17, 2002.

271 The beach cleanup was feckless: I covered the cleanup firsthand for more than a year for the *Anchorage Daily News.*

271 critics attacked the companies: Revkin, "Exxon-Led Group"; Darcy Frey, "How Green Is BP?", *New York Times Magazine*, December 8, 2002; Brad Foss, Associated Press, "Research Partnership Questioned," *Anchorage Daily News*, November 11, 2002.

271 A Greenpeace staffer in Alaska: Melanie Duchin.

272 the only unlimited carbon swallower: The oceans can dissolve immense but finite quantities of carbon as well, based on a ratio of carbon dioxide concentration in the atmosphere and in the ocean, but that issue does not affect my underlying point.

272 the earth would likely still experience: Interview with Ron Stouffer of the NOAA Geophysical Fluid Dynamics Lab, October 15, 2002.

272 Stabilizing the climate at a higher temperature: Ken Caldeira, Atul K. Jain, Martin I. Hoffert, "Climate Sensitivity Uncertainty and the Need for Energy without CO_2 Emission," *Science*, vol. 299, no. 5615 (March 28, 2003): 2052. Using midrange as-

sumptions and the same speed of emissions reductions would put climate equilibrium 4 degrees C warmer in 2150.

274 In 1963, a storm hit Barrow: This paragraph is based on Ronald D. Brunner et al., "An Arctic Disaster and Its Policy Implications," *Arctic*, in press, March 2003.

274 that Nuvuk used to be a mile farther out: *The Journal of Rochfort Maguire, 1852–1854*, vol. 2, ed. John Bockstoce (London: Hakluyt Society, 1988), pp. 378–79.

275 the Iñupiat lost some of the adaptability: I am indebted to Anne Jensen for this insight.

275 The U.S. Army Corps of Engineers studied the idea: Brunner et al., "Arctic Disaster," and personal communication.

275 "It's manageable": George LeVasseur, Alaska Department of Transportation and Public Facilities.

275 asked Ted Stevens for money to relocate: Liz Ruskin, "Consolidating Villages Gets Consideration," *Anchorage Daily News*, August 11, 2002.

275 Shishmaref villagers feared: Liz Ruskin, "Panel OKs Shishmaref Move Money," *Anchorage Daily News*, July 25, 2002.

277 one writer predicts social breakdown: Gelbspan, *Heat Is On*, p. 12.

277 another predicts a healthy boost: Steven Milloy, "Does Global Warming Really Matter?" *USA Today*, July 19, 2001.

277 sea level rose four to eight inches: Intergovernmental Panel on Climate Change, *Climate Change 2001: Impact, Adaptation, and Vulnerability*, eds. James J. McCarthy et al. (Cambridge: Cambridge University Press, 2001), p. 348.

277 "threatens skiing, foliage": Barry G. Rabe, *Greenhouse & Statehouse: The Evolving State Government Role in Climate Change* (Arlington: Pew Center on Global Climate Change, 2002), p. 18; www.pewclimate.org.

277 Many other U.S. states and local governments: David Appell, "Acting Locally: In Curbing Greenhouse Gas Emissions, States Go It Alone," *Scientific American* 288, no. 6 (June 2003): 20.

278 "Spaceship Earth": Ronald D. Brunner, "Science and the Climate Change Regime," *Policy Sciences* 34, no. 1 (2001): 1–33.

278 a ten-thousand-year-old science: Ronald D. Brunner, "Predictions and Policy Decisions," *Technological Forecasting and Social Change* 62, no. 1–2 (August–September 1999): 73–78.

280 The last interior Iñupiat village: This is the same incident mentioned in chapter 2, described in Charles Brower, *Fifty Years Below Zero: A Lifetime of Adventure in the Far North* (1942; repr., Fairbanks: University of Alaska Press, 1994).

280 The last traditional Kivgiq: Chris B. Wooley and Rex A. Okakok, "Kivgiq: A Celebration of Who We Are," presented to the 16th Annual Meeting of the Alaska Anthropological Association, March 3–4, 1989, Anchorage.

Acknowledgments

As a writer, I get to choose how to present myself and my ideas. My subjects do not, and that is what makes their participation—the decision to open up their lives and work—so generous and brave. That, and the time it takes. I often feel like Detective Columbo from the TV series of the same name, who irritates his suspects by showing up again and again at inopportune times: "One more thing, professor."

Savik Crew and Oliver Leavitt Crew welcomed me into the most important events of their year with the knowledge not only that I could be a pest but also that, as a writer, I could do real harm. I treasure the life-expanding experiences and friendships from that time. I am proud to know Richard Glenn and I owe to him the existence of this book. Likewise, I admire the members of the 2002 Arctic snow transect team too much to think of them simply as sources; after being allowed into their lives as a writer, I wish to stay as a friend. Others, including Craig George, Glenn Sheehan, Anne Jensen, Glenn Juday, Kenny Toovak, and Jim Maslanik, opened sensitive zones in their hearts and creative corridors in their minds that proved to be rich springs of meaning. They reminded me once again that many people, everywhere, think deeply about the world every day. Their ideas and Richard's are all over the book masquerading as my own.

The scientists at IARC and Toolik Field Camp showed me generosity and patience in explaining work that I frequently could not understand. Some may not like their portrayal, but they have my full respect. These are among the brightest people in society, and it is a reassuring commentary on humanity that, without great reward, they have dedicated themselves to this difficult work on our collective behalf.

Since I interviewed well over two hundred people and received assistance from more, it would be excessive to list everyone here individually. I have tried to make clear in the text and notes who helped; if I quoted someone, you can be sure that, at the least, I ate up plenty of his or her time. The notes and text also highlight some who were particularly generous. Many unnamed people helped as well, including a few who spent substantial time with me and were never mentioned for reasons of the story, including Bjartmar Sveinbjornsson of the University of Alaska Anchorage and Jim Randerson and his team from Caltech. In addition, several people read some or all of the manuscript in draft and made corrections or comments (although they are in no way responsible for the accuracy of the final product), including (in no particular order): Richard and Arlene Glenn, Henry Huntington, Henrik Wahren, Gerard Roe, Amanda Lynch, April Cheuvront, Bernie Zak, Craig George, David Cole, Ed Berg, Glenn Sheehan, Anne Jensen, Igor Polyakov, Jackie Grebmeier, Jana Harcharek, Jim Maslanik, John Walsh, Johnny Lin, Matthew Sturm, Mike Wallace, Ron Brunner, Syun-Ichi Akasofu, and Alden Todd.

Finally, I would simply be a cad if I did not mention my mother-in-law, Barbara Hill, and my mother, Caroline Wohlforth, without whose child care, family support, and child driving I would not have been able to research or write this book. They frequently put their lives on hold for us. My wife, Barbara, and our children, Robin, Julia, Joseph, and Becky, tolerated my absences and late nights and weekends in my study for two years; the box of broken toys on my shelf is overflowing, but the children do not complain. My editor, Becky Saletan, is brilliant and sensitive. She takes the time to truly understand the work before making improvements (a difficult and astonishingly rare skill). Roger Straus is a great man, as everyone knows, but he is also a good and a loyal one. Alička Pistek and Robert Meyerowitz helped me get started.

For the reader, the people in these pages are literary characters. For me, they are that and real people as well. For themselves, they are real people only, and the characters I have created are something else entirely. Yet they helped create them, and did so altruistically, to spread knowledge. As a writer, I carry that gift in trust for readers. In that sense, I hope the entire book is an acknowledgment, and that it does some good.

Index